Penguin Books
The Electronic Estate

Trevor Barr is Senior Lecturer in Media Studies at Swinburne Institute of Technology in Victoria.

He is well-known for his media commentaries on ABC radio and has contributed to the *Age*, *Nation Review*, *Arena*, *Australian Video and Communications* and *Look and Listen*. He is the author of *Reflections of Reality: The Media in Australia* (1977).

TREVOR BARR

THE ELECTRONIC ESTATE:

NEW COMMUNICATIONS MEDIA AND AUSTRALIA

Penguin Books

Penguin Books Australia Ltd,
487 Maroondah Highway, P.O. Box 257
Ringwood, Victoria, 3134, Australia
Penguin Books Ltd,
Harmondsworth, Middlesex, England
Penguin Books,
40 West 23rd Street, New York, N.Y. 10010, U.S.A.
Penguin Books Canada Ltd,
2801 John Street, Markham, Ontario, Canada L3R 1B4
Penguin Books (N.Z.) Ltd,
182-190 Wairau Road, Auckland 10, New Zealand

First published by Penguin Books Australia, 1985

Typeset in Plantin and Eurostyle by Leader Composition Pty Ltd

Made and printed in Australia
by The Dominion Press - Hedges & Bell, Victoria

CIP

Barr, Trevor, 1940-
The electronic estate.

Bibliography.
Includes index.
ISBN 0 14 008006 6.

1. Telecommunication systems - Social aspects - Australia. 2. Telecommunication policy - Australia.
I. Title.

384'.0994

Contents

Acknowledgements

The genesis of this book was an overseas study program in 1980 which showed me that the key issues in communications policy are international, and Australians share many common concerns about technological change and the future. I should like to thank those in the United States, Canada and Britain who so freely gave me their time and knowledge.

I am also indebted to many people in Australia who provided me with critical assessment of particular chapters of the draft manuscript: Roger Banks, Liza Dethridge, Stuart Fowler, Ashley Goldsworthy, David Lane, Alan Morgan, John Morgan, Peter Phillips, John O'Hara and Brian Skelley. My special gratitude to Jan White, who, with Helen Nolan, agonised over typing my copy. My thanks also to John Thom, who in a flash of inspiration gave me the title.

Several staff members of Penguin Books Australia deserve recognition for their contribution; publisher Brian Johns for accepting the concept, Jackie Yowell for her support, and especially John Curtain for his constructive editorial criticism and good humour.

My deepest thanks must go to the members of my family who tolerated both my travel overseas and my writing at home.

1 New Technology: Ambiguities of Abundance

No one announced the new Industrial Revolution. Workers in heavy industrial plants find that their new colleagues are computerised robots; secretaries are now planted in front of word processing machines; journalists feed copy into video display terminals; bank tellers instantaneously retrieve financial information from distant data banks; viewers call up on their television screens computerised data about the weather, share markets, gambling competitions, travel and news.

Despite a period of prolonged economic recession, information technology has become the fastest growing sector of production for many developed countries. Not only has the computer and information-technology industry boomed, but substantial development has also evolved in communications technologies – cable and radiated subscription television, communication satellites, video cassette recorders, videodiscs, videotex and teletext. These developments are more than technical and economic; more than a new cycle of investment. The computer, telecommunications and media industries are shaping new patterns of production and consumption central to social change.

The information revolution operates in cultures where corporate and individual insecurity is translated into perceived needs for fast, efficient information to gain competitive advantage. Transnational corporations have devoted astonishing entrepreneurial zeal to capitalise on such perceived needs as they shift the new technologies around the world.

No discussion about the outcomes for technology can be value free. Technology is embedded within a socio-political system and each social structure has choices to make about how, or whether, new technologies will be used. Technology is situated within society rather than apart from it, and questions about their interrelationship are essentially political.

Technology is too often presented as something which is external to society, with a dynamic of its own to which society can attribute blame for disruptive effects, to which society 'must learn to adapt'. Effective debate about new technology must be concerned with the power relationships within society and with value judgements about the shape and direction of that society.

In an economic context a significant trend in developed countries since the Second World War had seen agricultural and manufacturing sectors of employment progressively decline in favour of the tertiary or service sector. In more recent years a further trend has suggested the emergence of an 'information economy'.

It is now widely asserted that half of the total national income of the USA today is derived from the production, processing and distribution of information materials and services.[1] Moreover, the number of jobs created in this domain has, in recent years, exceeded the total number of jobs created in the agricultural, industrial and other service sectors.

Fritz Malchup, back in 1962, made brave assumptions when he claimed that knowledge production and distribution might, in an industrial society such as the USA, contribute 29 per cent of gross national product.[2] Later, Stanford University economist Marc Porat, in a nine-volume report, re-analysed economic data from the files of the US Department of Commerce from the perspective of information occupations and information-related activities. Porat defined a fourth economic sector as the information sector, in which he included all of the occupations where information processing was the dominant activity. So this sector not only included new information industries such as computer workers, but also teachers, lawyers, accountants, clerical workers and professionals. Porat, by essentially re-classifying some of the occupations of the old tertiary sector in a newly created fourth sector, the information sector, claimed that this contained at least 50 per cent of US employment. The term information sector is now widely used to comprise white-collar or non-production workers in the labour force.

Some economists define only two sectors in the work force, an information sector composed of people whose primary function is to create, process and handle information and a production sector composed of people whose primary function is to create, process and handle physical goods. By these criteria, there can be no doubt that there has been a decline in the production sector of advanced economies for both wealth and employment, and a growth in activities based on the collection, processing, storage and retrieval of information. Classification problems are inevitable. A foreman of a factory, for instance, may divide his or her time between doing administration work and manual work. On this classification, too, a sales manager of a car-production company now becomes an information worker. Some of these problems of classification have, no doubt, led to an inflation of the importance of the notion of an information society. But the development of a range of new industries based on high technologies and increased significance for information processing is obvious.

Information, as such, is a slippery concept. The term is used to cover so much – data, copy, intelligence and news, with other specialised connotations, are generally used in narrower senses than the all-embracing distinctive term, 'information'. Information can also be represented in many different interchangeable forms – as words, pictures, speech, writing, symbols or patterns. Furthermore, the gathering, evaluating, storing and transmitting of information content are performed by a wide range of different physical processes – pen and paper, printing presses, broadcasting channels and computers. The term, 'information', is used in such a sweeping manner that it is little wonder that economists have struggled to come to an understanding of the growth of the 'information economy' and then to refine their tools of analysis.

The growth of information technology as an industry has been fuelled within a culture of profound uncertainty. What is commonly expected of information in a business context is that its productive input will improve the efficiency of existing processes, or mobilise idle resources or open new markets. Information *per se* is expected somehow to generate revenue. Assumptions have developed that more information must be better than less; that when management widens the quantity of information available to them they will make better decisions; that an increase in information available leads to an increase in productivity.

Labels such as 'the information society' or 'the information revolution' have become highly fashionable. A Swedish parliamentarian has suggested that if Karl Marx was living now he would not have written *Das Kapital* but *Die Information.* American sociologist Daniel Bell, arguably the leading 'post industrial' theorist a decade ago, suggested that we were living in a 'knowledge society'. Bell claimed that 'if capital and labour are the main structural features of industrial society, information and knowledge are those of post industrial society.'[3] In the coming century, said Bell, the emergence of a new social framework based on telecommunications may be decisive for the way in which economic and social exchanges are conducted, the way knowledge is created and retrieved, and the character of occupations in which people engage. His post industrial society referred to a revolution in the organisation and processing of information and knowledge, in which computers played a central role. In the context of telecommunications, he stressed the perceived change from a goods-producing to a service economy, and the creation of a 'new intellectual technology' as key tools for systems analysis and decision theory.[4]

As microelectronics became more pervasive in the 1970s, and as more new information technologies emerged, social forecasters saw these developments as manifestations of Bell's post industrial or knowledge society. The 'wired society' or 'electronic utopia' was emerging, we were told, with information readily on tap in a new, profoundly humane industrial revolution. Alvin Toffler boldly asserted that robotics, artificial intelligence and sophisticated new computer software would allow businesses to specialise, decentralise, and 'respond effectively to individual consumer needs'.[5] For James Martin, ideologue of the wired society, it would be possible to build a world without pollution, without massive destruction of nature's beauty, without human drudgery.[6]

The computer industry spawned a new breed of technological utopians. For Peter Large, in his *The Micro Revolution,* 'this quiet revolution has produced a few horrors and many niggling annoyances, caused by inadequately planned computer systems and the scarcity of computer-skilled people' but 'microelectronics will create more employment opportunities, though different ones, as all technological advances have so far'.[7] Christopher Evans, in his *The Mighty Micro,* claimed that 'the new "unemployment", if we can use the dreaded word at all in this context, will be of quite different order, and much closer to a kind of affluent redundancy'.[8] The optimist's prize must, however, go to the more recent work of English professor Tom Stonier. Stonier attempts to comfort us with the profound assertions that 'just as the industrial economy eliminated slavery, famine and pestilence, so will the post industrial economy eliminate authoritarianism, war and strife. For the first time in history, the rate at which we solve problems will exceed the rate at which they appear. This will leave us to get on with the real business of the next century. To take care of each other.'[9]

Technological utopians have created myths for the post-industrial era. They have advanced the notion of the liberation of mankind from material constraints. It is a gigantic leap from the observation of the acknowledged inflation of the tertiary sector in contemporary Western economies to proposing a golden age of services where we are all somehow less dependent on tangible goods. Moreover, utopians believe that the wired society, with the development of inter-dependent information and mass communication networks, will foster mutual understanding and peace in a new internationally harmonious community. We are still waiting for Marshall McLuhan's global village for television twenty years after he first propounded the idea. International peace is clearly dependent on factors other than the sophistication of modern communication networks.

What seem to be ignored by this school of naivety are the power structures and struggles in contemporary societies. So many writers have equated developments in the field of information technology with a new kind of industrial revolution, where 'information' is somehow the new raw material which will solve political, economic, ecological and industrialisation problems. Yet who will be the controllers of the new society? Where is the drive coming from for information-technology growth? If we are in a stage of an information revolution, who are the revolutionaries?

The information society may well intensify the struggle for survival and strengthen the monopoly of economic activity over the social and political dimensions of our lives. The great danger with contemporary technological changes and the introduction of new communications technologies is that existing monopolists who own the technologies may strengthen their stranglehold. The most important yet least explored questions about the information society, this new electronic estate, deal with power relationships in society and their effects.

A classic text on the information society, *L'Informatisation de la Société,* has emerged from France. Simon Nora's and Alain Minc's neologism of *télématique* – the interconnections between telecommunications and computers – has become the established term to describe the French information program.[10] Nora's and Minc's unique study, commissioned by President Giscard d'Estaing in the late 1970s, tackled critical issues of telematics and national independence, cultural conflicts of a computerised society and the socialisation of information.

For Nora and Minc, the computer revolution was to have wide consequences affecting France's economic balance and its power relationships, and increasing the stakes of sovereignty. The gigantic American computer corporation, IBM, received special attention in that Nora and Minc argued that IBM was following a strategy that would enable it to set up a communication network, and to control it. When it did that, they warned, IBM would encroach upon a traditional sphere of government power: communications. They argued that policy must enhance the positive effects of telematics on productivity and that the challenge was building a system of connections that would allow information growth and social organisation to progress together. Their blueprint for development led to an innovative and ambitious telematics plan under the Mitterand government in the 1980s, though few international observers have been prepared to endorse it as a successful national strategy. Rather than muse about alleged benefits of a post-industrial society, the analysis of Nora and Minc got to the heart of the matter – balances of power and the new nexus in the relationship between corporate and state economic development.

The dynamism for astonishing growth in the electronics and communications industries is derived from the international organisation of capital and production. The 'take off' of information technologies can be largely attributed to trans-national corporations, whose international reach essentially determines which products will be distributed throughout the world.

A small group of wealthy trans-national corporations are primarily responsible for distributing new technologies around the globe. The major producers of electronics hardware are nearly all members of America's top 100 companies. Several of them, for over a decade, have consistently appeared in the top 20: IBM, General Electric, ITT, Western Electric (a division of the giant communications corporation, American Telegraph and Telephone - AT&T), and Westinghouse Electric.[11] The leap achieved by the major electronics corporations is crucial to an explanation of the dynamism of the information sector. At the beginning of the 1980s, *Fortune's* Directory of the 50 leading exporters in the USA showed that they amassed total sales of $57 billion for 1980, which accounted for more than a quarter of all US merchandise sold abroad.[12] Eleven corporations of the top 50 exporters were in information industries of some kind: in rank order, IBM (No. 9 of the top 50 exporters), Eastmann Kodak (No. 11), Raytheon (15), Hewlett-Packard (19), Rockwell International (35), Digital Equipment (38), Sperry (39), NCR (41), Emerson Electric (45), Control Data (47), and Minnesota, Mining and Manufacturing (48). Two of those companies, Digital Equipment and Control Data, are young companies who came from nowhere to leap into the bracket of top 50 exporters as superstar sunrise corporations.

Trans-national US-based corporations have long dominated international computer-equipment markets: in 1976 US computer-equipment firms accounted for 87% by value of the sale of the world's computers.[13] Increased foreign competition in the early 1980s has reduced their share to about three-quarters of the international market.[14] In the telecommunications field, an OECD report shows that 12 firms alone account for over 80% of world equipment sales, with three American corporations having a large slice of this $40 billion market annually.[15] In 1982, the top 10 US-based trans-national corporations together spent more than $13 billion on Research and Development, or more than the GNP of many nations.[16] Be it computers - mainframes, small business, minis or desk tops - or telecommunications equipment or word processors or videocassette recorders, a few, and often the same, big brand names tend to dominate the international markets. Australia has no members of its own in this big league - our information revolution has been imported.

The size and market dominance of the world's leading telecommunications corporate, AT&T, and the world's largest computer corporation, IBM, are absolutely staggering. Data compiled by the US Justice Department in 1978 for their anti-trust suit against the giant telecommunications monopolist showed that AT&T then enjoyed assets of over $US93 billion – AT&T is the world's largest corporation ever.[17] Their total annual revenues exceeded the budget of every state of the union and represented nearly 2% of the gross national product of the USA. A wholly owned subsidiary of AT&T, Western Electric, has consistently figured in the list of America's top 25 companies. In 1978, the Justice Department showed that AT&T's gross profits exceeded $12.4 million a day![18]

IBM's track record in the computer field shows an undisputed market leader. IBM has become the most valued stock of the Fortune 500 industrials in the 1980s. Its outstanding common shares in 1982 were worth more than twice as much as runner-up Exxon's.[19] Unprecedented demand for its computer products has enabled IBM to lift its earnings by about a third per annum during the 1980s, with about half of its sales normally coming from exports. It is almost totally dominant in smaller overseas markets and generally commands at least 50 per cent of total computer product sales in major countries. AT&T and IBM are so big, and reach so far, that they are more than just the senior members of America's multi-national club – combined, they virtually transcend capitalism itself.

American professor Herbert Schiller has argued that the growth of information technology rests on a strategy that ensures that new industries must emerge to fertilise or replace the old to ensure the continued massive private accumulation of capital.[20] The continued well being of America's information industries is dependent on world markets; information machinery, products and services, and 'know how' have become increasingly significant components of American foreign trade. For Schiller, the new information technologies, and the continued development of computerisation, will expand the influence of the already dominant elements of the social order in their assigned role of maintaining American world economic hegemony.

Schiller argues that the elevation of private corporate over public social interest is especially transparent in the dynamic communications-information sector. The driving force of the corporate policy makers of trans-nationals is to get advanced communications technology installed in as many places around the world as quickly as possible. Schiller claims that the changing division of labour internationally is being installed, guided and implemented largely by the trans-national corporate system. Whilst the justification for new communications technologies rests heavily on its 'promise to reduce inequalities and extend educational, cultural and human opportunities', the institutional arrangements that are being established, in Schiller's view, work towards an altogether different outcome. He warns that the catalyst for growth of information technology in contemporary economies – privatisation and commercialisation – is not influenced in the slightest by social need or larger public interest.

Complementary studies by French theorist Armand Mattelart have argued that the structural crisis which the world capitalist economies are undergoing signifies and demands a remodelling of the modes of communication.[21] His detailed research on the power and control of American based trans-national corporations concluded that cultural production and media around the world are subjected to a new phase of expansionary American imperialism.

Mattelart's work has been particularly valuable in showing how computerisation and new communications technologies owe their derivation to the military domain. He showed that it was the financial contribution of the US Defense Department which allowed the data-processing industry of the 1950s to take off. He argued that the sudden substantial increases in the US Department of Defense in the mid 1970s stimulated the electronics and aerospace industry; military contracts financed the development, research, test and evaluation programs then initiated by a few big corporations. This was an inevitable response to the malaise of American industry in the early 1970s when the American economy entered a crisis of over production with retreat from Vietnam and the end of a period of major purchases of space vehicles and commercial aircraft. Since the mid 1970s, according to Mattelart, we have essentially witnessed the civil application of those technologies whose exclusive military cycle is over. Whilst we have returned to a peacetime economy, the military dynamic has not lost its effect.

Mattelart argued that analysis of the progress of civil applications of satellite technology indicated that domestic satellites are actually the fall-out of a technology conceived for military purposes. He documented the staggering sums spent on the American space program and showed that NASA was often charged with specific defence projects. For instance, the first series of supposedly civil satellites, Syncom 1, 2 and 3, were built at the request of the Department of Defense, who subsequently directed them for use in military communications above the Pacific Ocean. RCA has, since the 1960s, built a system of meteorological defence satellites for the Pentagon. As the US has progressively got its own military requirements in place with space satellites, NASA has moved onto a new stage of trans-national entrepreneurialism by launching domestic satellites for many countries around the world. Australia becomes a member of the domestic satellite space club in 1985 by courtesy of NASA and American high technology.

For Mattelart, like Schiller, the growth of American trans-national corporations is a profound change in the relationships between the state and private-enterprise organisations. His claim is that the history of the relationship between the state and aerospace and electronics firms shows how industry can pressurise the bureaucratic apparatus to adopt its own criteria of modernisation. For Mattelart, the 'canalisation of defence and space technologies into the civil domain' precipitated the opening up of the state apparatus to the owners of high technology, who by default become social planners. Thus, claims Mattelart, a few private firms are promoted to the position of 'a technopolitical instrument'.

American high-technology trans-national corporations have thrived to such an extent that, by gaining world leadership in computers, electronics and telecommunications, they have contributed to the emergence of a new international economic order. Schiller and Mattelart reminded us that the producers of guided missiles, military satellites and sophisticated modern weapons are the same conglomerates who manufacture and market the electronic components of international information systems. The power of individual nation states has been subsumed by a small group of international corporate giants who dominate the economic growth of countries which are in transition towards a post-industrial society. The world's economic power balance is now held by this corporate group, who are essentially answerable only to themselves, and who promote and shape technological and socio-economic change, without knowing what will be the outcome of such processes.

In the mid 1980s, the USA faces intensifying international competition as Japan emerges to challenge American technological leadership. The Japanese government, through its Ministry of International Trade and Industry (MITI), has provided the centralised government planning and finance necessary to elevate the nation's leading electronic corporations into major participation in foreign high-technology investments. The Japanese strategy has been previously one of targeted concentration on long-term industrial security, rather than the usual pre-occupation with short term government economic fine tuning. Japan's success has been based on highly centralised planning, where goals were set for industrial development: heavy metal industries after World War II, then emphasis on car making and consumer electronics, and now the frontiers of microelectronics. The dynamism of the corporate sector has seen a sharp growth in the financial strength of Japanese corporations in foreign economies. According to a MITI white paper, Japan in 1984 was the fourth largest direct foreign investor of all major economies: Japan's average annual growth rate of direct foreign investment, for 1976-82, stood at 19% which was more than double the US growth rate of 8% for the same period.[22]

It is Japan that Americans most fear as their high technology competitor. Japan has come to compete head-on with American corporations, not only in the car industry and television-sets market, but also in the computer, telecommunications, videocassette-recorder and general electronics industry itself. Japan has now taken over from the USA in the development of large memory chips and is world leader in development of the 256k chip. In specialised communication fields such as fibre optics and computer graphics, Japan is the world's leading innovator.

Fujitsu, Japan's largest computer corporation, has now replaced IBM as the top seller of computers in Japan, a computer market second in size only to that of the USA.[23] The leading Japanese consumer-electronics firms such as Sony, Sanyo, Matshushita, Sharp and Casio, all key exporters, are well placed to profit from the opportunities presented as the data-processing and communication technologies converge. The major contemporary development in Japanese industry is research into a fifth-generation computer – a device able to communicate with people by sight, voice and touch. The Japanese Information Processing Development Centre (JIPDEL) is co-ordinating a 10-year research plan aimed to develop a fifth-generation system by 1990.

The USA and Japan, the leading competitors in electronics

innovation, are fierce rivals in a new global economic combat for domination of overseas markets in the information fields. The Australian information market is part of a new economic order, with greater internationalisation of capital and technology.

2 Australia's Information Economy: Technology meets Communication

Australia's electronic estate is composed of two interrelated economic sectors: the information-technology industry where computer hardware and software are central, and the communications industry, which is primarily the telecommunications and media service businesses, both private and public. While there are problems in compiling accurate data on the industries as a whole, or on individual industries, the trend is clear: information technology and the communications industry are now major components of the Australian economy, and both will become critical indicators in determining Australian wealth and economic stability in the future. Information industries as a whole have become the most significant investment area within the Australian economy. Indicators suggest that, apart from construction, information technology will directly account for about 20% of domestic capital expenditure during the next five years.[1] Investment in computer systems, hardware and software, on present indicators, could reasonably be estimated to account for $15 – $20 billion within the next five years. In addition, capital expenditure programs in communications-network building, by public and private organisations, could amount to $10 - $15 billion in the next five years. Telecom, for instance, the public monopoly terrestrial telecommunication carrier, planned to spend $1.92 billion on its capital works program for just the 1984-5 financial year, and Aussat Pty Ltd, the domestic satellite authority, has embarked upon an expansive capital program.[2] With multiplier effects, these kinds of investments collectively amount to a conservatively estimated $25 – $35 billion during the next five years. The leading area of economic development in the 1980s may not be the one predicted for mining, oil and gas projects at the beginning of the decade – the so-called resources boom – but the less heralded investments and economic development associated with Australia's information industries.

According to the Scott-Little report, information technology worldwide during the next decade will grow in accumulated revenue from US$365 billion (1982) to almost $800 billion in 1992, an 8% compound p.a. growth rate, and for the Australian market from about $A3.2 billion to $7.1 billion over the same period.[3] With Australia's economic subservience to foreign products and companies in information technology, a central question for Australia is who will benefit from the transformation taking place.

We must urgently develop and implement policies to ensure that the economic benefits and social engineering associated with Australia's information economy are managed by our own national collective hands.

There can be no doubt, in terms of computer communications hardware developments in recent years, that Australia's information revolution is imported. Trans-national computer corporations dominate the Australian market. It is most regrettable that reliable figures are not available to indicate the revenues of computer-hardware manufacturers who operate in Australia. A computer industry advisory group report to the Victorian government claimed that Australia spent $13.60 on imported computer equipment for every dollar spent on Australian manufactured equipment – the most unbalanced proportion of any advanced country.[4] IDC Research Australia claimed that the total information-processing market in Australia for 1982-3, including office systems and communications, was worth $3700 million.[5] IBM Australia's 1983 report shows an annual turnover of $520 million for the year ended December 1983, up 35% on the 1982 figure, and nearly five times more than its nearest competitor. In terms of information-processing-related revenues the leaders in Australia, after IBM, are ICL, NCR, Digital Equipment, Hewlett-Packard, Wang, Honeywell, Burroughs, Sperry, Control Data, Facom and Prime (not in order).

The only sizeable Australian owned enterprise in the electronics and telematics field is Amalgamated Wireless Australasia (AWA).

Opportunities for local product to date have essentially only been in the software and services markets. In information-technology hardware, Australia has been colonised.

Australia's domestic satellite system is foreign high-tech hardware at its best – launched from America's Cape Carnaveral by NASA; built by America's leading satellite corporation, Hughes International; with major earth stations supplied by Mitsubishi of Japan. Some concessions were built into the tendering system for 'local' manufacturers but these are largely subsidiary companies of trans-national corporations.

The videocassette market, which has boomed to such an extent that the industry claims that more than a quarter of Australian homes have a VCR, is another goldmine for foreign production. Japanese-manufactured hardware has led the world in VCRs with spectacular export growth from Matshushita and Sony. In Australia, Matshushita market through the brand name of National, who offer us Peter-Allen commercials calling Australia and the products which surround him, home. We manufacture no videocassette recorders in Australia; in production terms nothing could be less national than National.

There has been staggering growth of the service industries of Australia within the communications field in the past decade. Table 2.1 presents annual revenues for five branches of Australia's communications industry – the Postal Service, the telecommunications services revenue of Telecom, the Overseas Telecommunications Commission, commercial broadcasting (radio and television), and the national broadcasting services of the ABC. As a group, these industries have enjoyed aggregate increases of revenues from $1467 million in 1974 to $6200 million in 1982-3. In these components of Australia's communication industry alone, the revenue growth rate from 1974 to 1982-3 was a 323% increase, compared to an inflation growth rate of 122% on the CPI during the same period. The combined revenues of Australia Post, Telecom, OTC and the broadcasting sector, public and private, amounted to 2.6% of Australia's GDP in 1982-3. Apart from financial and banking services, this spectacular growth rate for Australia's communications industry is unmatched by any other industrial sector.

Top growth rate was recorded by commercial television, whose

TABLE 2.1 AUSTRALIAN COMMUNICATIONS INDUSTRY STRUCTURE: REVENUE

Branch	1950		1960		1965	
	$'000	%	$'000	%	$'000	%
Post						
Postage	27,893	30	79,935	25	103,032	21
Total	30,721	33	86,522	27	112,191	22
Telecom						
Telephone Rental	11,644	12	46,561	15	82,175	16
Telephone Calls	28,206	30	97,843	31	142,722	28
Telegrams	7,559	8	9,830	3	11,423	2
Leased Telegraph	494	1	3,515	1	5,846	1
Telex Rent & Calls						
Total	49,736	53	164,651	52	257,853	51
OTC						
Telephone	117	0.1	544	0.2	3,049	0.6
Telex			472	0.1	2,235	0.4
Telegraph	1,096	1.2	4,424	1.4	6,515	1.3
Leases			419	0.1	1,738	0.3
Total	1,389	1	5,979	2	14,034	3
Commercial Broadcasting						
Radio	6,357	7	20,503	6	24,550	5
TV			20,638	6	49,695	10
Total	6,357	7	41,141	13	74,245	15
National Broadcasting & TV	5,993	6	20,093	6	43,715	9
Grand Total	94,196	100	318,386	100	502,038	100

Updated from D. Lamberton, 'Structure and Growth of Communications Services' in K.A. Tucker (ed.), *Economics of the Australian Service Sector*, Croom Helm, London, 1977.

NOTES:

a. OTC review figures for 1950 and 1960 understate its share due to changes in the accounting system.

b. National Broadcasting and TV revenue figures are not meaningful and total expenditure including capital expenditure has been used as a surrogate variable.

c. Only the main items of revenue are shown in the breakdown.

d. Discrepancies in percentage totals due to rounding errors.

e. OTC figures are shown to one decimal place to illustrate comparative internal changes over time; no additional accuracy is implied.

1970		1974		1978-79		1982-83	
$'000	%	$'000	%	$'000	%	$'000	%
149,035	17	224,246	15	528,978	15	870,719	14
161,866	19	244,996	17	646,490	18	994,434	16
126,669	15	235,816	16	533,295	15	1,001,212	16
280,757	33	517,375	35	1,211,057	34	2,011,546	32
15,120	2	19,499	1	35,187		28,998	
9,253	1	15,155	1	7,221		10,156	
				46,385		91,304	
463,378	54	853,435	58	3,525,329	58	3,635,848	59
11,076	1.3	28,041	1.9	88,308		210,599	3
6,522	0.8	16,346	0.7	29,789		15,918	
9,821	1.1	9,760	0.7	9,315		8,349	
7,943	0.9	7,204	0.5	7,999		10,363	
38,231	4	66,074	4	143,215	4	300,072	4
37,059	4	56,587	4	125,627	3	221,329	3
91,191	11	132,299	9	407,621	11	816,048	13
128,250	15	188,886	13	533,248	15	1,037,377	17
63,009	7	113,766	8	157,972	4	232,931	4
854,734	100	1,467,157	100	3,525,329	100	6,200,662	100

f. National Broadcasting and TV is the ABC and does not include SBS or community broadcasting revenues.

g. Commercial Broadcasting and Television include ancillary revenues, e.g. 1983 – TV: $99 million; Radio: $6 million.

SOURCES: Postmaster General's Department Financial & Statistical Bulletin (Annual), Australian Broadcasting Control Board Annual Reports, Australia: Treasurer's Statement of Receipts and Expenditures accompanied by the Report of the Auditor General (Annual), Overseas Telecommunications Commission Annual Reports, Telecom Annual Reports, Australian Broadcasting Tribunal Reports.

revenues leapt from $132 million in 1974 to $816 million in 1982-3 – a 518% increase. Telecom's revenue performance is much more impressive than is indicated by a cursory glance. Whilst it was a fourfold growth for Telecom – from $853 million in 1974 to $3636 million in 1982-3 – telecommunications prices were actually reduced in real terms. During this period the consumer price index rose by 122% while the Telecom price index rose by 53%.[6] Some charges to consumers, such as long-distance trunk calls, were actually reduced. This pricing practice applies, of course, only to public bodies. OTC's pricing practice was similar in that its 1982 Annual Report stated that 'for the eighth year in succession, tariff reductions were introduced to a number of major overseas destinations in either or both the high customer-usage automatic telephone and telex services.'[7] Despite these cost concessions, OTC revenues grew from $66 million in 1974 to $301 million for the year ended 1983 – a 371 per cent increase – with an eightfold growth in its major service, international telephone calls. This kind of analysis begs questions about the adequacy of revenue as a criterion and how growth in the communications sector is measured. What, too, of other industries such as newspaper services, book publishing and printing, photographic services, motion picture and video services? Data is difficult to acquire for these industries. In print media, for instance, data on book publishing is inadequate; information from the Australian Book Publishers' Association covers only its membership. Newspaper-revenue data is not always isolated as a product line in company reports. Table 2.1 data could be constructed only because of the public ownership character of most of those organisations – they are either public statutory corporations or, as in the case of commercial broadcasting, accountable to a public body, the Australian Broadcasting Tribunal. The ABT tabulates aggregate annual broadcasting company data under the *Broadcasting and Television Act* (Section 106[1]), but this allows for a long delay in filing data, which explains the reliance on 1982-3 financial data in this section. Data about revenue from the computer industry, both hardware and software, is generally unavailable. Two public bodies, the Industries Assistance Commission and the Australian Law Reform Commission, have lamented in their reports the lack of statistical information about computer hardware and software and computer usage in Australia.[8] The IAC points out that the

Australian Bureau of Statistics *Survey of Manufacturing Activity* does not include such statistics, that there is no overseas trade data for computer software and that imported data for computer hardware 'has a number of problems'. The IAC Report on computer hardware and software stated that 'industry organisations presented only limited data on the operations of their members' and that 'although the Commission requested additional enterprise information from individual witnesses, little was received.'[9] This is a sad commentary on the closed nature of Australia's privatised computer industry. It is a supreme irony that, for the principal industry of the information age, computers, we simply cannot get adequate information.

The indicators available point to substantial economic growth in their field in general, yet we are unable to isolate detailed trends. Political parties claim that dealing with unemployment is a top social priority, yet data about where people are employed and new industry prospects is totally inadequate. Whatever happened to Barry Jones's idea in *Sleepers Wake!* that Australia 'establish a National Labour Force Planning Commission to work with trade unions, employers, states, schools and tertiary educational institutions, local government and Commonwealth authorities to collect and analyse data about changing employment patterns and expectations'?[10]

One of the most confusing outcomes of the growth of an information economy is the divergence of positions regarding the likely effects of microelectronics on employment numbers. In Australia in the mid 1980s there is an unsolved technological-employment equation, where jobs are disappearing in some information industries yet expanding in others. In some areas of work it appears that new electronic systems require less labour than did their mechanical predecessors, and that the new information technologies often reduce the range of skills that workers require to perform their jobs. For example, introduction of digital telephone exchanges, computerised typesetting and word processors appears to be associated with labour shedding, though some findings are contradictory. There are examples of deskilling in the printing industry and with telecommunications-equipment installation, and the trend towards an automated office has displaced some of the more skilled clerical jobs. It is difficult to predict the overall aggregate employment effect associated with the adoption of new information technology. In Australia's indigenous printing industry at the beginning of the 1980s, for instance, more than 10,000 jobs were lost.[11] On the other hand, however, in the Australian domestic computer software industry, the number of enterprises has increased from 100 in 1974, to 400 in 1978, to 900 in 1983, for an estimated total employment in computer-software houses in Australia of 6000 persons. [12]

Misconceptions exist about the future job prospects of a high-technology society. What ought to be challenged is the notion that future economic growth in Australia is solely dependent on nurturing new high-technology industries. There is a tendency to elevate technology as an isolated economic factor, and to suggest that technological growth, by itself, can somehow generate new wealth, with associated increases in job prospects. The pattern for developed economies in recent years is one of limited annual economic growth, without concomitant increases in employment, a phenomenon known as 'jobless growth'.

TABLE 2.2 AUSTRALIAN COMMUNICATIONS INDUSTRY STRUCTURE: EMPLOYMENT ANNUAL REVENUE AND FULL TIME STAFF

	1978-79		1982-83	
	$'000	Staff	$'000	Staff
Australia Post	646,490	39,156	994,434	32,459
Telecom	3,525,329	87,440	3,635,848	88,095
OTC	143,215	2,116	300,072	2,612
ABC	157,972	6,251	232,931	6,431
	4,473,006	134,963	5,163,285	129,597

SOURCES: Annual Reports of Australia Post, Telecom, OTC and the Australian Broadcasting Commission.

The revenue data for the telecommunications industry in Table 2.1 can be compared with employment data for those organisations which are public bodies (Table 2.2). Table 2.2 shows an aggregate growth in revenue for those four branches from $4.47 billion in 1978-9 to $5.16 billion in 1982-3 accompanied by a net employment loss of 5356 jobs. No comparable employment data is available for Australia's commercial broadcasting industries.

It is now assumed that more people in Australia are employed in the collection, processing, storage and dissemination of information than work in farming, mining and manufacturing combined. Barry Jones has pointed out that between 1965 and 1982 there were 2,060,000 new jobs created in Australia: yet 'not one of those new jobs *net* was in manufacturing, in fact the sector lost 150,000 in that period (-7.3 per cent).'[13] Services, Jones added, is by far the dominant employer in Australia, with the service sector of the economy contributing 67.2% in 1981, highlighting 'the paradigm shift in our economy away from raw materials and muscle power towards knowledge and skill.'[14] Future employment prospects are likely to be in the service sector, with new jobs more associated with fast food, health, gambling, property security, beauty, home improvements, travel, relaxation and hobbies. But the service sector appears to be moving towards low-wage or part-time jobs with low skill requirements.

Issues of employment also involve the nature of work and the impact that the new technological work place has on workers. A body of evidence indicates that the indiscriminate introduction of some new technologies has had serious effects on the health and welfare of working people. The common repetitive strain injury is tenosynovitis, which refers to inflammation of tendons and the tendon sheath in the wrist and hand, but there are numerous other injuries encompassed by the term 'occupational overuse syndrome' which also gives rise to prolonged pain. According to Workers' Compensation Commission statistics in NSW, the reported incidence of synovitis, bursitis and tenosynovitis in women trebled in the years 1978-82.[15] In a comprehensive study undertaken on repetitive strain injury (RSI) in Australia, a consultant's report undertaken for Telecom in Western Australia suggested that one in five of a sample of 418 telephone operators in computerised exchanges had suffered serious RSI disability since the equipment was installed in 1980.[16] In News Ltd's Sydney office there were 30 reported cases of Australian Journalists' Association members suffering from RSI.[17]

From the evidence, a general case can be made that new information technology has been introduced in such a haphazard manner that management has not been aware of, or has failed to meet its obligations towards, the protection of the occupational health of employees. Assuming that the hardware will not be banished, the problems can be tackled in two ways. Employers and employees could co-operate to adopt, as required, job rotation, frequent breaks, a slower pace of work, ergonomic improvements, acclimatisation at slower rates, early reporting of symptoms, medical surveillance of possible sufferers and rehabilitation of victims.[18] A second course of action is that Australia may have to follow the example of countries such as West Germany and Japan, where tough occupational health legislation had to be introduced to protect workers from overuse injury.

Underpinning these issues involved in employment and occupational health are the questions of why there has been so little understanding and debate of the social impact of technological change.

Jurgen Habermas, German academic and notable philosopher of the post-industrial age, poses many questions about contemporary technological development and the institutional framework of society.[19] For Habermas, when the state finds it imperative to follow scientific technical progress which can guarantee economic growth and stability, society must be run by technical experts who perceive problems as technical problems which can be solved by technical expertise. A problem arises because the social system does not obey the logic of scientific progress – values and interests clash, social choices arise and complex questions need to be reflected upon. Increasingly, though, argues Haberman, technical developments with far reaching social consequences are accepted without question and implementation is aridly concerned with expertise, 'not reflection'.

The processes of open and thorough public discussion about new technology in Australian society, to provide the basis for rational decision-making by government, have been inadequate. Ostensibly there has been long investigation about technological change and the role of new technologies in Australian society. Following the Myers Report of the Committee of Inquiry into Technological Change, we have had a plethora of Commonwealth government inquiries into particular new technologies and existing communications institutions.[20] The most notable are the national task-force satellite inquiry (1978), the Australian Broadcasting Tribunal inquiry into the possible introduction of cable and subscription television (1982), the Davidson inquiry into Telecom and telecommunications services (1982), the Dix inquiry into the ABC (1981), and the Bradley inquiry into the Australian Postal Commission (1982).[21]

These processes of public inquiry have been unsatisfactory on several grounds. They essentially decided and recommend on means and techniques - on how to introduce new technologies, not whether they are desirable or beneficial. The inquiries must be seen in the context of a ruling-class ideology which equates technological change with technological progress. In such a climate, how could a group of technocrats and bureaucrats appointed by government possibly recommend against, for instance, the introduction of a domestic satellite? Inquiries face a minefield of competing interests and, whilst genuine attempts are made to stand above particular interests, they have been unable to initiate social investigation which is independent of submissions presented and witnesses who offer advice.[22].

Incoming presentations are subjected to a ratings game - the bigger the company, or the more influential the politician, the greater the ritual with which the advice is tendered to the inquiry. The assumptions are commonly that the introduction of a new technology provides a service which must inevitably meet needs. The equation always begins with the technology - no inquiry began with a thorough investigation of needs from which possible courses of action flow, part of which may involve other new technologies.

Committees of inquiry must be seen to be in touch with the 'real' world - hence complex issues such as employment implications, deskilling, alienation, isolation, privacy and social inequalities are secondary in consideration to whether 'we let things rip'. In any case, with the short time these inquiries were given to consider their findings and present their report, the complex issues cannot become matters for long-term analysis or reflection.

Most inquiries were framed in a way that assumed that the technologies were discrete, separate, compartmentalised entities. This is inappropriate to technologies which are increasingly converging and where, even in terms of technical considerations, investigation needs to cut across several technologies. Those who framed the terms of reference for different inquiries must have been dismayed to find that overlaps were impossible to avoid, but astonished that, on some major communication-policy issues, different inquiries presented directly contradictory findings.

The need for rational debate and open public discussion about technology is not filled by our mass media. This is not only because the established economies of the commercial media require them to deliver viewers to advertisers through popular programs. Television in Australia has rarely been used to explore complex subject matter; and vexed issues relating to technological change in society have not made it onto the agenda of concerns, as defined by television, in either private or public organisations.

In the few cases where matters about technology are canvassed on television, programs are usually structured with a beginning, a middle and an end. We are rarely presented with dissent or unresolved questions; a set of accepted conventions of presentation has emerged which provides the viewers with ease of comprehension and slides over the complexity of the subject matter.

Technological issues, like politics, are personalised in the mass media. Human interest 'angles' must be found in covering any topic. The accepted conventions of presentation have contributed to Barry Jones having the highest public profile of any Minister for Science (and formerly also for Technology) in any Australian government. Public-affairs television is now essentially about mood and visual impressions at the expense of substance. The coverage of technology in Australian society on television has presented much about the personality quirks of Barry Jones – we have seen likeable Barry as former Quiz whiz and as collector of autographs of famous people, and we have learnt about his eating habits and dislike of pubs.[23] This is the level of discourse at which technological matters are often communicated on television. The Minister for the Future, as he has described himself, has not been given a forum or conventions of presentation which allow him to debate issues of fundamental importance to Australian society.

Other technological topics offered in public affairs programs are usually dealt with in terms of admiration for gee-whiz gadgetry, or an inventor made good, or as part of the high-tech dream of the future. Commercials invariably present an assumed nexus for new technology of: greater efficiency = increased productivity = better standard of living. Militaristic symbols are often invoked in commercials, effectively playing on aggressive notions so well exploited by the genre of science fiction films. Or they play on insecurities where anachronistic oldies, out of touch with computers, are shown desperately searching to learn about the machines which the kids born into the information age so readily understand

and enjoy. Commercials about new technology are consistently about gross forms of competition and are generally antagonistic towards humanistic culture.

The development of the information society has been technologically driven, with concentration heavily on the means and techniques of distribution for ever-increasing quantities of information. The new forms of information hardware are presumed to be about advancement and progress. They offer, for instance, quicker ways of paying bills by electronic funds transfer, more efficient teleshopping with supermarket specials called up on screen via videotex, or advanced learning at home via the computer terminal. Yet in human terms the provision of greater services does not necessarily mean the satisfaction of needs. The development of more efficient information distribution systems is inevitable, but will we find telebanking, teleshopping and electronic mail more satisfying? Will any ordinary mortals ever be asked whether they want these electronic changes? New technologies ought to be means to ends, and the ends need to be debated and understood before we embark on change. Is the present spate of technological changes compatible with the social order that the Australian community wants?

In broadcasting, we have embarked on an expensive and technologically sophisticated national domestic-satellite system which will give greater efficiency of signal distribution, allow more live broadcasting and encourage more channels to be made available. Yet, will the programs be any better or different? Will Australia's age of satellite television still continue the abysmal neglect of television audiences? The domestic satellite was sold politically on the promise of greater equity of services and to make available broadcasting to the outback. Yet who are these outback communities, black or white, and what do we know about them economically, socially and culturally? What kind of programs do they want or need? Will they ever be asked? Will Aborigines be fed a diet of *Yes Minister, The National, Brideshead Revisited* and *Australia You're Standing In It*? Will people of the outback be able to make their own programs?

We have embarked upon profound changes in methods of communication, yet have not bothered to find out about the nature of the communities that these technologies are intended to serve. Media and telecommunication developments are concentrating so much on the conduits of delivery – cable, microwave, radiated subscription television, satellite, fibre optic – that the fundamental questions of what is to be delivered, and why, have not only gone unanswered but they have not been asked.

Ideally, these technologies will foster greater understanding between human beings and provide richer sources of satisfaction in the way people live their lives. Speed, efficiency of delivery and retrieval, are presumed to be valuable attributes of the new technologies. Television news services now promote the notion of providing the latest, faster technologies to interpret 'this ever more complex world'. Yet speed has never been a prerequisite for effective learning, and rarely a component of enjoyment. It simply does not follow that more information automatically facilitates greater understanding, or that faster delivery of information provides better communications. The technical capacity to assemble, process and deliver information has expanded exponentially, but the human capacity to absorb and interpret information cannot expand commensurately. A further paradox of the information age is that whilst there are considerable technical possibilities for greater communication, the essentially closed and exclusive nature of broadcasting and telecommunications systems could mean more limited opportunities for widespread involvement and participation. The 1980s will undoubtedly see substantial increases in the quantity of information, but there are no guarantees that the quality or diversity of content will be improved, or that public accessibility will be encouraged. We may be building a society with more information available than ever before, but which actually has less knowledge and which is less communicative.

3 Governments and Technology: Directed Economies or Market Forces

Governments in developed countries have long been involved with media, information and communication industries. Freedom of the press is in uneasy stages of resolution around the world, but the newspapers of democracies are commercially operated, widely free from direct government intervention. The electronic media of radio and television often evolved as a dual system: on one side a public broadcasting organisation dependent on government funding, yet with programming independence; on the other a commercial sector subjected to government regulations.

In most developed countries, the telephone network is government-owned or ultimately government-controlled, with the most notable exception, the USA. Postal systems are universally government-owned with their functions often delegated to a national statutory corporation. However, advertising markets, and information-processing markets, including all components of the computer industry, have widely developed as private and unregulated business markets. What has emerged in recent years on the agenda of the politics of the future in most developed countries is a reassessment of the traditional roles of government in technological change, choice and opportunity.

The leading economies of Western democracies have embarked upon strategies for growth in the communication industries with a basic objective to build an efficient and dynamic information economy. High priority has been given to the development of a strong telecommunications infra-structure, generally composed of the terrestrial telecommunications network, domestic communication satellites, new outlets for broadcasting, a computer service industry, and associated industries which provide hardware and software products for the telecommunication and broadcasting systems. A host of vexed issues arise in policy formulation related to information economies of the 1980s. Two broad contrasting strategies have emerged: on the one hand, the freewheeling leave-it-to private enterprise approach, where the speed and form of technological change is fashioned by marketplace forces; on the other hand a strategy where the government assesses technological options, and plans or directs the nature of change. These are not absolutely exclusive choices, but most countries can be located on a continuum from laissez-faire economies to centrally planned, government-directed information policies. Australian governments have, in recent years, grappled with the problem of where they stand on such a communications-policy continuum. The role of government as a regulator in telecommunications and broadcasting is becoming increasingly complex as a result of the convergence of the technologies. Whereas, in the past, print media, broadcasting and telecommunications were essentially compartmentalised as separate industries, the contemporary technological developments are moving towards the integration of formerly disparate technologies. The speed and degree of technical sophistication of new communications technologies, and their convergence, will pose monumental problems for policy makers and regulators. In the past, three major regulatory structures have been used in the media-communications field in Australia.

First, print media emerged free from government regulation; Australian governments have no formal control over newspaper, book and magazine publishers. No licences are required to establish companies in print media and apart from legal sanctions such as libel, newspaper, magazine and book publishers are free to publish as they choose.

Second, in broadcasting, Australia opted for the dual system. There is a small group of private broadcasters (who were licensed by government for radio in the 1930s and 1940s and, essentially the same group, for television in the 1950s and 1960s), and the publicly funded broadcasting authority, the Australian Broadcasting Corporation (the ABC), formerly a Commission. The conventional explanation of the need for government to regulate broadcasters is that limitations of spectrum space necessitated a process of selection by licensing. However, a more convincing explanation for greater government regulation in broadcasting than print media involves the notion of economic protectionism for a small group of entrenched broadcasters. Clearly in the past there could have been many more operators in the broadcasting system. How this might be changed by the introduction of new technologies begs questions about future communications power structures in Australian society.

Third, there is the common-carrier model for telecommunications, where the Australian government established a public monopoly which has exclusive rights to deliver particular telecommunications signals – Telecom for the national telecommunications terrestrial network, and OTC for international telecommunications services via Australia. Whether the telecommunications services should continue to be provided on a public monopoly basis, or opened up to competition, involves important economic questions not only for Australia but for other countries also in the next decade.

There are several pressures, of different kinds, on these three established models for communication policy: unregulated print, licensed broadcasters and public common-carrier telecommunications monopolies. The computer industry, though, has developed in a laissez-faire marketplace, with virtually no technical, content or regulatory standards. With the convergence of the technologies, the lines of demarcation have become difficult to establish between unregulated computer services and regulated telecommunications services.

Is, for instance, a service that enhances a transmission line (with data-switching or processing capabilities) a traditional telecommunications activity or a new computer offering? The merging of print with new electronic media also raises some new regulatory questions. Is a newspaper delivered by teletext to the home television set merely an extension of print media and thus as 'free' of regulation as any other newspaper, or does it then become a broadcast medium, subject to regulation? Are the inter-active services, such as telebanking or teleshopping, made possible in cable television systems through the television set, the domain of broadcasters or common carriers? In the USA and Europe today regulators grapple with the issue of whether direct-broadcast satellite services (DBS) are 'broadcasting' or new hybrid services. Regulators struggle to relate new services to old categories. The growth of electronic media systems is eroding the old regulatory and legal foundations.

Underpinning the reassessment of regulatory and communications public policy questions is the broader issue – the extent to which government ought to be the architect of the information economy in controlling the mix of public and private-sector involvement. Many industrialised countries, including Australia, have been in the process of examining whether they should introduce greater competition in their broadcasting and telecommunications industries.

There are a number of reasons for this. Briefly, during the period of prolonged economic recession, private-sector companies have been forced to critically examine their whole array of operational costs. Communication costs within the company, especially telephone costs, have been analysed and cheaper alternatives are obviously keenly sought. There is also a view within the private sector that new forms of business-communication services ought to be handled by private companies who are generally perceived to be more responsive to their needs than suppliers from a public corporation. How, argue the private-sector protagonists for change, can a single public organisation effectively respond to companies with complex business-communication needs to adopt increasingly sophisticated computer, office and information systems?

Some companies simply want a piece of the action. They are keen to diversify into the high-growth computer-communications area. What is the justification, they ask, for leaving entrepreneurial ventures to monopolies staffed by bureaucrats? Many companies who missed out on the original licences want to be part of a new private-enterprise shakedown which cable and satellite technologies might bring. The catch cry of private enterprise in telematics is to deregulate the market place. The general assumption, rarely challenged, is that greater competition will lead to economic growth, greater efficiency, and a better range of services for the community.

Deregulation of communication industries must be set in the wider context of overall national economic strategy. Towards the end of the 1970s, Milton Friedman's anti-Keynesian financial strategy of monetarism was widely embraced by Western democratic governments. Its essence was encapsulated in President

Reagan's 1980 campaign slogan of 'best government is the least'. Monetarism sought to reduce levels of public expenditure and asserted that all forms of state intervention should be reduced in the interests of free market operation. For Friedman, economic stability can only come through a private-sector-led recovery. Central to this form of economic management in the US and UK has been a move towards greater de-bureaucratisation and deregulation. The Carter and Reagan administrations deregulated the railways, the airlines, the banks and energy controls. In telematics, the Reagan administration dropped two long-fought anti-trust suits against American Telephone & Telegraph (AT&T) and IBM – gestures which symbolised a competitive free-for-all of unprecedented magnitude. The resolute Thatcher administration has embarked on 're-privatisation', or selling back to private enterprise some parts of nationalised industries. In telecommunications, the British government has argued that it is better to be 'driven forward by the forces of competition' and it has set in motion the privatisation of British Telecom.[1] Similarly, the Japanese government, early in 1984, decided on the privatisation of the country's telecom monopoly, Nippon Telegraph and Telephone Corporation, from April 1985. This will open a Pandora's box for intensely profit-oriented companies of many countries, including US corporations, many of whom have long wanted access to some of the monopoly's vast services.

In Australia, the Fraser government (1975-83) embraced the rhetoric, but not the reality, of monetarism and small government. It actually increased the number of federal statutory corporations during its term of office, and handed its successor a record level of federal government deficit.[2] The Liberal-National Party coalition widely espoused a deregulation philosophy, but implemented few policies that would bring about a freer market economy in Australia. For instance, it reviewed the banking system during its last three years of office but did not implement the deregulation policy which its own review had recommended. There was a considerable body of opinion from the business community, and within its own ranks, which argued for loosening business regulation. John Elliot, Managing Director of Elders IXL, had argued that 'regulatory intervention delays decision making, creates uncertainty, saps enterprise initiative, hampers innovation and is an unnecessary and costly waste of time.'[3] For Elliot, free-enterprise sentinel and member of the Victorian Liberal Party executive, the imposed process of regulation lost opportunities to innovate and grow. Politicians, he pleaded, should 'recognise that the market mechanism works, that it allows people to demonstrate their own capacity to show initiative and enterprise.' For Elliot, 'governments should help foster individuals and companies to get things done, not continually tell them why they are not allowed to do them.'[4] The question of whether private enterprise in unregulated markets 'gets things done' is an issue to which we shall later return.

The rhetoric of the Australia Labor Party has long been directed towards remedying the injustices and inequalities of capitalism. Its platform stressed the objectives of redistribution of political and economic power, equal access and rights to employment, education, information and technology as part of social justice and equality for individuals. A stated party objective is support for a competitive private sector, controlled and owned by Australians, which operates within clear social guidelines. Members of the first Hawke Cabinet tended to argue the case for a buoyant mixed economy, in preference to elevating the role of the private sector in development strategies. There were few statements of position from the first Hawke government on greater privatisation. The strongest rejection of the concept came, rather oddly, from John Dawkins, Minister for Finance. Speaking at the National Press Club in March 1984, he stated that 'we reject completely the notion . . . that public enterprise is unproductive and makes no contribution to wealth and growth . . . Those who would argue for privatisation are blinkered by short term self interest.'[5] His statement of intent on behalf of the Labor government represented a fundamentally different position from that of Liberal Party strategists. According to Dawkins, Labor held that 'the public sector has a major role to play in the lives of all Australians, that it is a force for social and economic advancement, that without it Australia would be a poorer place, the gap between rich and poor greatly accentuated and the services and facilities now available to all would be provided only in a most haphazard, arbitrary, inefficient and uneconomical way.'[6]

These broadly stated philosophical stances of Australia's two major political parties on the issue of priorities for the public vs private sector, and the degree of acceptance of the international trend towards greater privatisation, are crucial for an understanding of the different positions taken by the Fraser and Hawke governments on telematics policy.

In telecommunications, the Fraser administration had intended to embark on a strategy for greater privatisation wherever possible, but it achieved little. The Fraser government established a national inquiry into Telecom, known as the Davidson inquiry, with its first term of reference as clear indication of its desire for greater privatisation – 'the extent to which the private sector could be more widely involved in the provision of existing or proposed telecommunications services in Australia.'[7] Actually, whilst the Davidson report found some new opportunities for private companies in telecommunications markets, it recommended a stronger role for the public authority in terms of Telecom's management of the major telecommunications networks. Whilst some senior members of the Hawke government had wavered before the beginning of the 1983 election campaign, Labor in office opted strongly for the retention of Telecom's public monopoly for terrestrial services. We shall see later how the issue of privatisation was crucial to the Fraser government's acceptance of the proposal that Australia needed its own domestic satellite system.

The Fraser government denied Telecom's plan in 1981 to be the national common carrier for videotex services, a decision which was widely condemned by videotex interest groups within the private sector, and one which was reversed by the Hawke government in 1983. The key issue was whether to give a public-sector organisation common-carrier status for the introduction of a new communications technology. In postal services, Sir Phillip Lynch's ill-fated 'Razor Gang' attempted to initiate greater privatisation of the Australian Post Office with their recommendation to disband the courier service of Australia Post. Curiously, the Fraser-appointed Bradley inquiry, whose task was to consider greater privatisation of postal services, later recommended re-establishment of the courier service and retention of the monopoly of the Australian Postal Commission! The Hawke government supported this position in 1983.

While, in practice, the Fraser government forays into privatisation failed in these areas of telecommunications and communications services, there are those who will ensure that the debate continues.

Models for Communication Policy: The USA and France

The development of communication policy in Australian society ought to be seen in the context of international trends. It is useful to contrast the diametrically opposed strategies of the USA and France. The USA represents the archcapitalist model, with lack of centralised planning, limited government involvement, deregulated marketplaces and new services evolving according to entrepreneurial zeal and the financial strength of private corporations. France, though culturally different and much smaller than the USA, provides a model for technological development which is ideologically at the other extreme. The plan to convert France into an information economy is based on a centralised, nationalistic, planned model, with huge investments from the public sector designed to engineer social change.

United States' Communications Policy

Technological developments, it is argued in the USA, are fostered by private-profit maximisation, economic efficiency and consumer sovereignty; America leads the world in information technology because the private sector has been given unfettered opportunities to vigorously market new technologies, free from government interference. Such argument ignores to some extent the history of communications technology in the USA. The US government affected the evolution of the computer industry by creating an assured demand for its products. In early 1954, the government was the only major purchaser of computers; by 1962 the government market still represented about one-half of total computer sales; in 1982 US government purchases accounted for 80 per cent of 'super computers'.[8] The US computer industry owes its development in part to large post-war government-sponsored defence and aerospace projects, which created a burgeoning demand for computer hardware and data-processing capabilities. Similarly, America's leadership in communications satellites and associated industries is derived from ambitious government-sponsored space programs. In the 1980s, however, the government's role has been conveniently forgotten and the buzz word is deregulation.

In the field of communications, no contemporary US administration has ever seriously engaged in systematic national planning, with successive governments remaining passive and acquiescing to the power of oligopolistic industries. The US government has progressively abandoned its limited presence in domestic communications policy development. While the Carter administration consolidated many federal programs in a single National Information and Telecommunications Administration (NITA), this did not lead to the development of a consolidated national policy. About eighteen federal agencies, together with the independent regulatory body, the Federal Communications Commission (FCC), remain involved with policy and operational matters. The Reagan administration has no major bureaucratic organisation, such as a Department of Communications, to advise, develop and co-ordinate policies on telematics. The federal agency responsible for broadcasting regulation during the last half century, the FCC, has progressively divested itself of authority in favour of deregulation. In fact in recent years in the US, deregulation has progressively moved towards unregulation in many areas.

These policies are an extension of US communications history in which policy decisions have been fashioned by competition and confrontation. Congress has made only minor amendments in recent years to the 1934 Communications Act, and telecommunications and broadcasting policy has essentially been developed by the rule-making of the FCC, and decisions of the courts. In the US, the judicial system, rather than the elected Congress, has long been the more potent force in shaping communications policy. In broadcasting, the legal philosophy which underpinned the establishment of broadcasting – public allocation and management of limited spectrum space – has progressively had less relevance with the introduction of new technologies of spectrum abundance.

American broadcasting has long been dominated by a few major commercial networks. Today, RCA Inc. (the parent company of NBC), CBS Inc., and American Broadcasting Companies Inc., are major American corporations. Each has its roots in broadcasting: NBC started its first radio network in 1926 and CBS began in 1927. Companies which controlled the radio networks initially determined the make-up of television networking; CBS and NBC in effect extended their radio networks into television networks. ABC lagged in size until the 1960s, when it reached competitive parity. Today, these three big networks have a stranglehold on American television. The original rationale for licence allocations, in both radio and television, was to take care of local community needs, but in the US the commercial networks soon developed national sources of programming. There has never been an equivalent in the USA broadcasting system to the British Broadcasting Corporation (BBC) or the Australian Broadcasting Corporation (ABC), which developed as influential public broadcasting institutions financed respectively by direct listener-sponsor licence fees and through direct government appropriation. Public broadcasting in the US has tended to remain small, localised and sporadic in character, with many public radio and television stations educationally based. The Corporation for Public Broadcasting, set up in 1967, which provided a national inter-connection amongst the country's educational television stations, was not established with guaranteed assurances of public funding. With notable exceptions, the sad irony of the immensely wealthy US broadcasting system is the poverty of the public programming sector.

In telecommunications, the US developed a unique model. Unlike other countries, where the Ministry of Posts and Telecommunications, or an equivalent, presides over a public telecommunications monopoly corporation, the USA chose a private model. Telegraph services have been provided largely by what became America's largest corporation, AT&T. It was regulated by state public-utility commissions, which governed intra-state services and tariffs, and by the FCC, which regulated interstate tariffs and services. The US has had to face vexed regulatory questions associated with new telecommunications. Two decades ago communication technologies tended to remain isolated within particular organisations: telephone services belonged to AT&T, telegraph to Western Union, letters to the US Postal Service, and broadcasting was dominated by the three commercial networks, ABC, CBS and NBC. New technologies created new competition for the established order – satellites competed with telephony, cable with aerial broadcasting – and demanded a new regulatory framework. The political mood has been to let the new technologies develop, allow them to compete with the existing order, and hope that market forces can resolve the problems. The Reagan administration has embraced the deregulatory mood more than any previous American administration. Its stated position is that the role of government in the marketplace should be drastically shrunk in the hope that competition will best serve consumer interests by increasing efficiency, stimulating innovation and thereby producing the widest possible product choices. Two long-standing anti-trust suits against major corporations, AT&T and IBM, had remained in separate areas of the market by a consent decree. Put simply, IBM dominated two-thirds of the computer market, and AT&T's Bell system sold primarily voice-telephone services and had no direct competition. The Bell system, together with the independent telephone companies who are dependent on Bell's long lines, accounted for more than 95 per cent of telecommunications revenues. When IBM linked its computers to telecommunications, it used a telephone company, and when a Bell phone company wanted a computer it was serviced by IBM, or a competitor. Telecommunications and computer organisations were essentially separated until 1982. With the convergence of new technologies both corporations saw the need for new strategies – AT&T saw great prospects in computer-based data networking, and IBM wanted to diversify into communications fields other than computers.

In 1956 AT&T had entered into a consent decree not to enter the computer business in return for guarantee of its telephone monopoly business. The Nixon administration initiated anti-trust legislation in 1974 against AT&T, the world's largest non-government corporation, with protracted charges of anti-competitive activity. In the 1982 agreement, AT&T agreed with the US Justice Department to divest itself of its local telephone companies (more than 80% of America's telephone local-service business) but retained its long-lines business, the manufacturing organisation Western Electric, and the highly innovative research and development division, Bell Labs, which is best known for its invention of the transistor.[9] The Reagan administration deal allowed AT&T into unregulated fields, such as computers and data communications, for the first time. Essentially, this unleashed a highly formidable new competitor on US domestic *and* world markets. In the past, AT&T was also prohibited from the international market which provided telex, telegraph and specialised data-transmission services to overseas points. In December 1982, however, the FCC ruled to allow AT&T to enter the international data market.[10] The IBM anti-trust case lasted longer than did AT&T's – for 13 years the US Justice Department charged that IBM monopolised, or attempted to monopolise, the market for general-purpose electronic digital computers. With the case against IBM dismissed in 1982, the corporation was free to operate in the US as it had always done. Both the AT&T and IBM megajudgements show the extent to which deregulation has been accepted by the US administration. The stage is set for the world's largest telecommunications company, AT&T, to battle on any ground with the world's biggest computer corporation, IBM, for the burgeoning markets of information-processing systems.

Successive Chairmen of the FCC, under the Carter and Reagan administrations, have both stressed the importance of freedom of the marketplace. Chairman Charles Ferris in April 1980 said that 'the supply of communications products and services will be limited only by the ingenuity of businessmen and scientists' and that 'governments will no longer be a barrier that prevents or delays the introduction of innovations in technology'.[11] His replacement under the Reagan administration, Mark Fowler, followed with similar rhetoric in 1982 – 'the public interest is measured by the success or failure of broadcasters in the marketplace against each other, and against other competing technologies . . . government meddling, however well meant, must be avoided.'[12]

In the past decade, the American courts and the FCC have progressively removed most of the regulations that had previously stifled the growth of cable-television networks. This was a crucial factor in the rush at the beginning of the 1980s to win cable franchises in regions which still remained unwired. In 1984, some 34 million American homes were on a cable system, or 40% of all television homes.[13] The FCC also lifted its restrictions on pay television, attempting to foster the growth of pay-television networks which were delivered via cable systems, or delivered over the air by UHF stations (STV), or microwave multi-point distribution services (MDS). The FCC also proposed in 1980 the creation of a new class of community-oriented stations, low-powered UHF stations, and were swamped with thousands of applications from those who wished to start such stations. And, in the area of direct-broadcast satellites, the FCC had authorised by late 1982 eight companies to enter the DBS business which would involve in total the launch of 21 satellites to offer 43 channels of national service.[14] Meanwhile sales of videocassette recorders in the USA reached a record peak of an estimated seven million units sold during 1984. In terms of the range of communications hardware and services available, American society has become during this heightened phase of deregulation an electronic cornucopia.

Government played a key role in the initial development of the American computer industry with its purchasing policies, and it also stimulated the growth of communications-satellite technology; but in the mid 1980s the US government essentially remained a passive observer of technological change and development.

French Communications Policy

The French model for technological development provides stark contrast with that of the USA, as the most strongly centralised public-communications model amongst Western democracies. The radical Mitterrand socialist government, elected in May 1981, declared its strong commitment to building French society into an information economy as quickly as possible. Determined to reconcile economic recovery with economic equity, it adopted centralised state planning of some key components of the French private sector. Takeovers of specified private firms by government early in 1982 gave it more than half of the nation's informatics industry. The key objective of the socialist economic program was to revive French industry by nationalisation of specified future dynamic sectors. The strategy was to use public capital to boost investment dramatically, which in turn would stimulate domestic markets and build exports. The Mitterrand nationalisations were central to the attempt to move the French economy to a position of world competitiveness.

The strategy of a public model reinforces France's entrenched view of the fundamental importance of nationalism. State intervention has long characterised French industry. Under de Gaulle's post-war coalition government of 1944-47 coal, gas and electricity were nationalised, together with more than 50% of banking and insurance, and some major private manufactures were reorganised into large state-owned firms such as auto production (Renault) and civil aviation (Air France). Before the war, French broadcasting was organised as a dual system of private stations financed by advertising revenue, together with public stations financed by taxes, under the authority of the Ministry of Posts and Telecommunications. In 1944, after liberation, leaders who had emerged from the resistance wanted to end the wealth and power of private interest in mass media, and French broadcasting became a state administration as a branch of the Ministry of Information. François Mitterrand was the Minister of Information in the late 1940s.

Whilst there have been several changes to the organisational structure of broadcasting since the 1940s, there has always been a consensus in France in favour of 'statism', and government control of broadcasting to protect what was considered to be the best interests of France. In France, the state owns and controls all radio and television broadcasting, though it delegates its powers over programs to companies who are given a large measure of autonomy. The French state also owns and controls the major French newsagency, *Agence France Presse,* and is the major shareholder in France's largest advertising agency. This state-directed media model also characterises the telecommunications industry, where the state agency, the *Postes, Télécommunications et Télédiffusion* (PTT) provides the publicly owned infra-structure upon which communications developments have been built.

Contemporary development of the French information society has been facilitated by the attention they have long given to military hardware. Under de Gaulle's administration, plans were developed to build their own arsenal of nuclear weapons, submarines, aircraft and missile technologies to reverse French dependence on the American military hegemony. The French public purse has funded the Mirage, Concorde and Airbus technologies, Aerospatiale helicopters, the Ariane space rocket for communications-satellite launches in competition with America's NASA, nuclear-breeder reactions, and nuclear testing in the Pacific. The present attempt to diversify into informatics actually benefits from the pool of skilled 'military' workers already available and from an established infra-structure of high-technology supplier companies.

It was under President Valéry Giscard d'Estaing that Simon Nora and Alain Minc, aided by a team of technical specialists, surveyed the computerisation of French society and proposed a unified national public policy to deal with the impending changes. Their report discussed telematics – the increasing interconnection between computers and telecommunications – in the context of French societal strain caused by continuous deficit in the national trade balance, the weakening of traditional internal consumer demand and growing French unemployment. They argued that only a deliberate policy of social change could both 'solve the problems raised by telematics and utilise its potential'.[15] Telematics, warned Nora and Minc, changed the stakes of sovereignty. IBM was given special attention in their report, described as a company which 'had played the multi-national game more intelligently than any other'.[16] They argued that lack of governmental action created a void 'quickly filled by the dynamism of IBM' which 'has everything it needs to become one of the world's great regulatory systems'.[17] They also argued that 'the risk remains that AT&T might reach an accord with IBM concerning the American and world markets, or even an agreement to share power in the world market alone, despite the lively competition they might give one another in the US'.[18] The alternative danger, according to Nora and Minc, was that France would use this new technology to reinforce mechanisms of rigidity, authority and domination. The challenge lies in the difficulty of building systems that will allow information and social organisation to progress together.

Under Mitterrand, the French embarked on one of the most ambitious nationally integrated communication plans in the world. With telematics, the service sector was expected to jump in productivity to achieve the kinds of gains enjoyed by agriculture and industry during the past twenty years. Computerisation was assumed to make a new type of economic growth both possible and necessary. The French plan was to use computer-communications technologies to construct an electronic digital infra-structure which would create mass markets for the services which these new technologies can provide. The Mitterrand government set out to implement the Nora and Minc strategy of creating a modern French information society by providing universal access to a comprehensive range of telecommunications facilities. Their program included the development of a public inter-active videotex service eventually for all of France, and the ultimate conversion of the printed telephone directory to an electronic directory as part of a substantial upgrading of their telecommunications infra-structure. These changes, especially the introduction of a nationwide public videotex system, were seen as public ventures of the 1980s analagous in importance to the development of railways in the late nineteenth century. These new communication 'highways' were seen as the key tools of participation, consultation and decentralisation of French *télématique* society.

According to France's Minister for Research and Industry, Jean-Pierre Chevenement, a strong showing in the electronics industry was fundamental to the long-term future of the French economy. He supervised a $20 billion five-year plan, designed to build France into the world's third leading technological nation after the USA and Japan.[19] A central part of the strategy has been to remove the duplication of French industrial research programs by forming a united group of powerful high-technology companies. Soon after its election, the Mitterrand government nationalised four of the largest French electronics companies – CII-Honeywell, Thomson – CSF/Thomson-Brandt Group, the Alcatel Electronique Div. of Compagnie Général d'Electricité and the Matra Group.[20] All, except Alcatel Electronique, were making losses, and major product breakthroughs and new markets were desperately needed. Government pinned its hope on the nationalised conglomerate to arrest its perpetual trade deficit of $1 billion annually in consumer electronics.

The three market sectors assigned top priority to capture home markets and develop a strong export base were electronic components, small computers and consumer electronics. Specific roles were assigned to these four information-technology corporations as part of French strategic planning. CII-Honeywell bear sole responsibility for all computers from micros to mainframes; the Thomson Group concentrate on consumer electronics; Alcatel handle office automation; whilst Matra and Thomson share the task of developing a viable semi-conductor industry.[21] This highly directed program of development is derivative of Japan's Ministry of International Trade and Industry (MITI) in that it divides the markets for particular high technologies amongst four designated companies.

For the French, the initial stage of the information society lies in the *télématique* mix of telecommunications and data processing, with videotex as the chief delivery system. Their Télématique system resulted from a 1975 policy decision to overhaul the notorious French telephone system about which the standard joke claimed that half the country waited for a telephone to be installed whilst the other half waited for the dial tone. In redesigning the telecommunications network, the plan was to build new electronic systems at a cost which would allow ready access for everyone. The first experimental application was general home videotex, the '3V' Teletel trial near Paris, while the second was the electronic directory service. The inter-active videotex system has an associated plastic card, called 'smart card', which contains a microprocessor and memory for use in electronic funds transfer. France's two largest credit-card organisations, *Carte Bleue* and *Crédit Agricole,* agreed on the distribution and promotion of a single payment card. The electronic directory service aims to provide telephone subscribers with better access to directory information, at no additional cost the the subscribers, but with savings for French Telecom in terms of paper directories and directory inquiry operators. The directory service allows users to consult a wide range of up-to-date information stored in computerised data bases through videotex terminals. With these two services subscribers have on-line access to a vast range of information, together with allied applications such as teleshopping, telebanking and videotex messaging. The two experimental applications, the 3V Teletel home videotex trial which ran from June 1981 to December 1982, and the electronic directory trials from summer 1980 to the end of 1982, convinced the French government that there was strong demand for both services. They claim that videotex services will experience faster growth than did the telephone, and their plan is to install three million videotex terminals in subscriber homes and offices by 1986.[22] The carrier for Télématique is the national public telecommunications utility, the PTT, but the data bases and customer services carried by the system are separated from the state-owned common carrier.

Few democracies have attempted national technological reconstruction through public-sector stimulus on such a scale as the French. For the Mitterrand strategists, high technology plans had to be of an all-embracing character. The French talk of *'filière électronique'*, which refers to the totality of an industry, including areas from other technologies and other skills which have to be brought to bear if industrial reconstruction is to succeed.

The conventional American business free-market philosophy is that government need not plan, or be directly involved in, a development strategy and that regulation is always a last recourse. The accepted French approach is fundamentally the opposite –

that the development of a French information society will only emerge from government-directed policies which include nationalisation of select companies within the private sector.

There are, however, some broad underlying similarities. The present communications revolution in both countries is a spin-off from their earlier industrial military aerospace ventures. Their mutual development of missile and weapons products provided an industrial infra-structure and scientific pool onto which the subsequent developments of computer and telecommunications industries could be grafted. Both governments, too, guaranteed that the initial stages of electronics and new information technologies were fostered by government demand. IBM in the 1950s and 1960s counted almost 60 per cent of revenues from American government contracts, and de Gaulle used government purchase by French companies as a strategy to reverse French dependence on American military technology. (The role of the American government as the prime sponsor is often ignored in discussions of its free-enterprise paradise.) There is general bi-partisan support in both countries for their new directions which avoids major shifts in policy orientation, as well as any possible industrial upheaval due to a change in government; this gives the benefits of a long-term continuity of strategic economic policy.

Both models have their share of critics. Put simply, the critics of the French system argue against the inherent waste and inefficiency of a bureaucratic model in their claim that the new information business must be highly entrepreneurial to take off. What is the motivation for drastic change within government, they ask, and how can such a dynamic new business be run from the desk of a ministry? The French system, add some critics, lacks marketplace consumer accountability and is replaced by a government opting for some new technologies and rejecting others. Governments, claim critics, are hopeless at picking industrial winners.

There are at least two major areas of criticism of American computer-communications policy. One school of thought within the US argues that privatisation does not go far enough – that record budget deficits, large defence programs and failure to reduce the costs of bureaucracy soak up funds and engineering skills which are critical to American innovation and competitiveness in new high-technology fields. A different school of US critics argues that those who promote, shape and intend to control the forces of technological and socio-economic change do not know what will be the outcome for the community as a whole. For them, America has unashamedly embarked on a winner-take-all strategy, with prizes for the select group of captains of capitalism. How, they ask, does such an unqualified private model ensure that economic and technological resource development will be compatible with desirable social goals? The same basic issues must now be related to developments in technology and communication policy in Australia.

4 The New Industrial State: Sunset and Sunrise

Large sections of Australia's traditional manufacturing base have substantially declined in output in recent years: the extent of factory closures and labour shedding has become so serious that Australia is now in a phase of de-industrialisation. A realisation has emerged that if Australia is to maintain its traditionally high standard of living, then what is needed is innovative economic and industrial planning, fresh trade initiatives, and a range of new industries to rebuild a moribund manufacturing base. Rarely challenged is the 'drive towards growth' economic model, though a prolonged period of economic attrition may lead to a radical re-assessment of work, employment and life-style policies. Whilst we remain with the present assumptions about conventional political economy, high technology has assumed a special role in projected economic developments. Naively in some quarters, high technology has become a panacea to remedy our economic ills, where traditional manufacturing industries are phased out for quick replacement by exciting new specialised technologies to secure Australia's long-term industrial future. So much is expected of high technology in the future – job creation, the revitalisation of Australian manufacturing, the way to increase international market competitiveness, and the vehicle by which acceptable standards of living can be maintained. The enormous task of reshaping Australia's industrial base must not be underestimated. The success of reconstruction will depend on the attitudes of people in government, business, science and education; on rethinking trade, tariff, marketing and procurement policies; and on bold investment practices in both the public and the private sector.

Australians have long ignored the problems of an outdated

industry structure: the manufacturing advantages of technological change have essentially been realised by overseas suppliers. Australia's exports of technological products have fallen so rapidly during the past decade that she is amongst the lowest technology-exporters of all members of the OECD. In 1982, Australia's exports of technological products per head of population dropped to a record low of US$73 per head of population; the OECD average was US$503 a head - only Iceland, Turkey and Greece were worse placed than Australia.[1] More specifically, in information-technology fields, of the estimated $3000 million worth of computers installed in Australia by 1983, it appears that more than 90% of domestic hardware supplies have been imported.[2] In telecommunications equipment, Australia has the worst export record amongst OECD countries with proportionally the highest trade deficit in the group.[3] The corollary is a substantial job loss in the telecommunications equipment field: employment between 1972 and 1978 fell from 30,490 to 20,882 jobs, a fall of 31 per cent compared with a reduction of 13 per cent for the overall manufacturing sector for that period.[4] One report estimated that the price Australia will pay for neglect in the microelectronics field, if present trends continue, will be a microelectronics equipment trade deficit of US$3.7 billion by 1988.[5] Australia has become progressively more dependent on imports for the supply of manufactured goods in the computer and telecommunications fields; we have become a lucrative outpost for many overseas interests.

A large majority of the companies involved with the supply of information technologies within Australia are overseas-owned. The major hardware components of the telematics market - computer, satellite and telecommunications equipment - are dominated by imports from multi-national companies whose activities are based in the USA, Europe and Japan. Whilst Telecom has generally bought about 90 per cent of its equipment from 'local' manufacturers, the companies are actually subsidiaries in Australia of multi-national parent companies. The major companies supplying the Australian market for telecommunications, satellite, computer and office equipment include L. M. Ericsson (Sweden), Siemens (West Germany), Hughes (USA), NEC (Japan), Plessey (UK), Rank Xerox (USA), IBM (USA), Mitsubishi (Japan), Philips (Holland), Olivetti (Italy), STC (USA). This international league of suppliers continues to benefit from the growing markets within Australia for computer hardware and software, a range of telecommunications equipment, from ordinary telephones to electronic telephone exchanges, office machines such as word processors, and the new developments in domestic-satellite equipment. There is, in fact, only one substantial Australian owned company in the electronics and telecommunications manufacturing field - Amalgamated Wireless (Australasia) Ltd, AWA.

There are several reasons why a predominantly overseas-owned and controlled manufacturing base must be fundamentally unacceptable to Australians. Inevitably, Australian manufacturing will remain essentially derivative, content to merely remain assemblers of products which were designed overseas. Australia has failed to develop its own industrial strategies, and failed to nurture its own inventors, researchers, technological innovators and entrepreneurs. By contrast, other small countries, such as Sweden, have developed an independent manufacturing exportable economy. Subsidiaries of trans-nationals in Australia export little from Australia in general, and they virtually ignore Australian based research and development. Overseas-owned companies generally have a disincentive to local investment, and their investment in research is home-based and hence at the expense of a substantial contribution to Australian R & D. There are, too, the traditional concerns as expressed by ASTEC, that 'in war or other emergency, Australia could be cut off from overseas supply of vital microelectronic devices' and that 'Australia must have in operation at least some basic production facilities.'[6] Comparatively little attention has been given by Australian governments to tackling this fundamental structural problem of the Australian economy. One school of reconstruction thought, led by Barry Jones, Labor Minister for Science, is that Australia must overcome the essential fragility of its industrial infra-structure and build its own exports, based on high-value advanced-technology goods. Various reasons have been proposed for our high-technology super-dependence on overseas enterprises - the unwillingness of past Commonwealth governments to act to create a base of Australian owned enterprises, the lack of local private-sector venture capital, the lack of natural advantages which blessed areas such as America's famous Silicon Valley, the high costs and associated risks of research and development required for modern technology, the economies of scale which favour international corporations, and Australia's failure to develop a dynamic intellectual scientific base. Only in recent years have these factors begun to be analysed and have constructive policies for change shown signs of emergence.

In the context of information technology, or telematics, the prospects for successful indigenous ventures have long been promising. A study of innovation in this field shows that the problem of development has not been lack of Australian inventiveness, but a long-term failure to convert technological innovations into successful entrepreneurial ventures which become owned and controlled by Australian companies. Our track record in communications shows some remarkable achievements in terms of inventiveness and scientific leadership. Radio science, for instance, involved pioneering work by Australian scientists. As McCracken has pointed out, our expertise in microwave technology, as developed by Christiansen and others, and in UHF and microwave astronomy in general, was as good as the best elsewhere in the world.[7] The first NASA deep-space tracking dish drew heavily on the experience gained in the construction of the Parkes radio-astronomy antenna, built in 1961. Australia had installed two Intelsat earth stations for international satellite communication by the end of the 1960s. We were involved in the beginnings of the computer industry: CSIRAC in 1949, SILLIAC in 1956, special-purpose digital telemetry systems in Woomera, and quick adoption of early generation mainframe computers. Australia has had the pioneers and scientific expertise in communications, yet failed to nurture indigenous companies as the technologies developed into billion-dollar markets.

It is not widely known that the old Weapons Research Establishment in South Australia made a vital contribution to the invention of xerography. What is widely known is how handsomely the process of photocopying was developed by Xerox in the USA. The basic patents of xerography are held in the names of officers of the old Weapons Research Establishment, at Elizabeth, in South Australia.[8] Xerography – what a technological fish to let off the hook!

Telecom's Research Laboratories have been responsible for some significant telecommunications inventions which have either been taken up by trans-national corporations or have lapsed in terms of Australian developmental possibilities. In her definitive history of Telecom, *Clear Across Australia,* Ann Moyal cites two examples of lost opportunity.[9] In 1972-73 Telecom research laboratories were in the vanguard of work overseas on stored program control (SPC) electronic exchanges, and it was then a real prospect for the next generation of switching equipment to be developed and manufactured in Australia. The 1977 Telecom commission, However, opted instead for L. M. Ericsson's SPC exchange system, called AXE. Ericsson have also acknowledged the contribution made by their Australian subsidiary company to parent operations in Stockholm in the automatic call distributor (ACD) system which evolved from the Australian PMG Research Laboratories. It was L. M. Ericsson, a Swedish company, which gained the full benefits of the worldwide marketing of ACD.

There is still evidence in the 1980s of Australia continuing to give away its inventions and enterprise in the communications field. For example, in recent years the Telecom Research Laboratories have developed a new product, the Digital Radio Concentrator System (DRCS), designed to meet the need for high-quality, reliable, moderately priced telephone services in outback communities. Telecom researched and developed this new product to the outline design stage, but they were then unable to interest any local manufacturers in its further development and marketing. In fact, Telecom handed over the innovation to NEC of Japan, who gratefully accepted this product of research from an Australian public corporation. The fate of the DRCS highlights Australia's failure to successfully develop the products of creative research and original design in specialised technological fields. In the three main stages of telecommunications product development – invention, then the engineering and commercialisation phase, followed by marketing – it is the second stage of indigenous commercial product development and manufacture in which Australia is acutely lagging.

What successful high-technology countries all possess is an infra-structure into which new manufactured products can be readily absorbed. Whereas one-off inventions tend to get 'lost' in technologically underdeveloped countries, innovations emerge naturally out of, and find a place in, a dynamic national technological infra-structure. The important point about the highly prized Silicon Valley model is not merely those advantages so commonly attributed to it, such as proximity to universities and government incentives, but the inter-dependence of high-technology industries upon each other. The French equivalent is *'filière électronique'*. Countries with a dynamic technological infra-structure today have achieved this from a long history of nurturing scientific talent and from a political and commercial commitment to the development of indigenous science-based industries.

The essential Australian high-technology development problem has been the vacuum which has long surrounded our scientific innovators. Hence, what is needed is not merely to encourage isolated techniques development, or foster a series of unrelated sunrise industries, but a bi-partisan commitment to develop a strategy which encompasses the broadest political, scientific and cultural development of a technological infra-structure which is Australian owned and controlled. A change in government in the USA, France or Japan would not fundamentally alter long-established policy, but in Australia, if past experience is any guide, a change of government could lead to profound changes in directions in these areas.

What must be clearly understood is that if Australians want to try to develop a high-technology base of their own, then there must be substantial national government involvement in planning. The development of a high-technology base in countries such as Sweden, Canada, Western Germany, France, and the United Kingdom, and even to some extent in the USA, has been dependent, in varied ways, on planned goals established by government to encourage their local technological industries. Surely in Australia now, the issue of whether government ought to be heavily involved as the architect of high-technology developments is no longer an open question. The issues are more about the forms and extent of that involvement and support.

A government-planned industrial model must not be confused with labels that are so often loosely substituted for it such as government interference or intervention or regulation. What is needed is for government to isolate key areas of economic development and to develop policies which will ensure that these designated sectors are given preferential treatment and full encouragement to grow and prosper. This is *not* an argument for propping up inefficient and costly industries with fiscal measures designed by government bureaucrats or keen regulators. Any government measures must have long-term goals of economic independence and viability of those industries designated for selective treatment. That does not mean that governments should merely tinker with tariffs, bounties and quotas in the hope that fiscal measures will remedy temporary anomalies. It means that a fuller range of government policies - research spending, tax incentives, export assistance, educational planning, community involvement schemes, venture capital policies and government procurement policies - must be analysed in the context of the key economic areas designated for high priority.

Information Technology as a Key Economic Sector

Australia must not merely find some new commercially viable industries, but develop a range of industries which collectively provide assured national economic benefit. The information-technology components of the economy ought to be designated as a key sector for economic development. The case for giving preferential government treatment to information technology is strong on many grounds.

First, information technology, or the computer-communications industries, have become progressively more imbedded in the Australian economy and culture in recent years. To isolate information technology as a key sector for economic preference is not then simply a hopeful sunrise-industry policy argument, or technological Tattslotto in that we hope that somehow one day we strike a new industrial jackpot by buying a lot of tickets. Support for information technology need not be a policy premised on technological determinism; rather, an extension of existing technological development within Australian society.

Second, information technology offers prospects for Australia to redress the balance of its present economic colonialism in high technology.

Third, economic industrial reconstruction in Australia must be based on capital-intensive industries; internationally uncompetitive labour costs in Australia rule out a strategy based on labour-intensive industrial redevelopment.

Fourth, the communications field does involve the knowledge, skills and techniques which produce goods and services for which there is an economic and social demand in Australia. Information-based industries, unlike smokestack industries, show all the signs of long-term permanent survival within Australia.

Some broad principles need to be established for key support of information technology in Australia.

A strategy of preferential treatment for information technology needs short-term and long-term goals. In the short term we ought to support the development of local products for which markets have been identified. Rather than casting a wide net in an R & D fishing expedition, we should home in on products which can expect results in the short term. What Australia cannot expect to do is to compete with major international competitors in high-volume manufactured products. Australia is unlikely ever to be a producer of mainframe computers in head-to-head competition with IBM. What we need to concentrate on is specialised manufacturing activities, designed to meet local requirements, at home and overseas, and funding specific support applications for which an identified demand exists. Examples will be given later, in the computer industry, telecommunications equipment, satellite technology and videotex hardware.

A long-term strategy ought to involve the export of information-technology products. No one should have illusions about the difficulties. Australian manufacturers of new high-technology products face major problems in developing reliable and viable export markets. Their labour costs per employee, and production costs per unit, are generally comparatively higher than their competitors. In marketing, the manufacturers of high-technology products have no equivalent to the Australian Wheat Board, where the producers can exchange their products for cash. Hence, they are forced either to set up their own distribution outlets in other countries or offer unrealistic high discounts to local prospective distributors. They also face the problem of assistance in that their counterpart producers overseas receive preferential treatment from their own governments, who long ago established procurement policies which favoured their own local industries.

The reality may be that there are few high-technology products to be manufactured in Australia which can be successfully marketed overseas, but we should be able to isolate some niche in export markets. We have already seen the sad history of Australia's failure in marketing to capitalise on its own inventiveness in the field of communications. What is required, as Barry Jones came to understand, is not simply R & D, but what he called R, D, Q & M – research, development, quality and marketing. There is little point investing major resources in product development in such a highly competitive area unless the intention is to produce high-quality products which are vigorously marketed.

As part of a national strategy to create a suitable high-technology infra-structure in Australia, with interrelated business, financial, educational and technical components, information technology has a key role to play. Several industries within the umbrella term, information technology – the computer industry, telecommunications manufacturing, satellite hardware and videotex equipment – need an integrated policy strategy to ensure that their development fosters indigenous talent in the best interests of Australia. Part of this strategy ought to be an attempt to revive flagging local manufacturing industries. Government support must be directed towards interrelated information-technology industries as part of a newly created manufacturing base which eventually becomes essentially Australian owned and controlled. If the development of information industries is to be as great as is widely predicted in the next decade, a benchmark policy decision is whether Australia initiates positive strategies which are economically nationalistic, or whether we allow ourselves to be exploited and colonised yet again.

What then are the possibilities? The hardware and software sections of the Australian computer market, despite some significant overlaps, have tended to develop as separate activities, with the local software section as the larger component enjoying a higher growth rate over the past few years. An Industries Assistance Commission report in 1984 on the computer industry, one of the few public investigations of what might be Australia's most promising sunrise industry, could only offer broad estimates of industrial patterns within the industry.[10] Their report offered no aggregate data for the revenue generated by the computer industry in Australia. It argued that the hardware sections of the market – mainframes, minicomputers, microcomputers and peripherals – were dominated by imports from large multinational companies, that 'there is no precise information on production and imports', but 'most witnesses were of the view that imports accounted for more than 90 per cent of domestic hardware supplies.'[11] The Australian Computer Equipment Suppliers' Association Ltd had suggested that computer hardware in Australia had grown at an average annual rate of about 20 per cent over the past five years.[12] According to the suppliers' association, employment in the local equipment industry had declined slightly in recent years, to an estimated total of 400 persons in 1983.

There is no local production of mainframe computers, but there has been some development in recent years of a number of companies who manufacture mini and microcomputers. The local industry, according to the IAC, supplies some specialised hardware requirements, especially terminal equipment. Local manufacturers supply the bulk of domestic requirements for wagering terminals and other computer-based totalisator equipment. In fact, one company, ATL Pty Ltd, has a healthy export market which accounts for 80 per cent of its production. The hardware industry also supplies a good portion of the markets for educational microcomputers, and some types of small-business accounting systems and cathode-ray tubes. But no accurate data was available regarding the extent to which exclusively Australian owned and controlled computer-hardware companies were involved with these developments.

The software component of the industry has experienced more substantial growth in recent years. According to estimates given to the IAC, the number of enterprises in the domestic software industry increased from 100 in 1974, to 400 in 1978, to 900 in 1983; aggregate turnover was estimated at $300 million, with employment at about 6000 persons.[13] This represents a pedominantly small-business development where software activities employ an estimated average of seven persons per enterprise. There is a high turnover of enterprises in this field: many Australian companies have fallen by the wayside through lack of financial resources, as against the long lead time before the product is fully developed and able to return revenue. Most enterprises in Australia develop a limited number of types of software for specialised usage. The division between hardware and software becomes blurred as more companies provide complete computer systems. For example, David Hartley Computer Australia Pty Ltd combine their own hardware and software into a complete system. It was David Hartley who distinguished himself with an impassioned keynote address to the Ninth Australian Computer conference, in Hobart in 1982, saying that, unless a national approach to the development of high technology was adopted, Australia was in danger of becoming 'the great white trash of the South Pacific'.[14] Other companies, aware that imported products are often unsuited to local demands, have concentrated on specialised niche markets; Moldflow's specific type of computer-aided plastics engineering and Applied Technology's educational systems for use in schools are examples. Clearly there are considerable opportunities, particularly in the software area, for the development of high-quality domestic computer products.

The IAC investigation into computer hardware and software provided a forum for companies and professional associations to express their views about the kinds of measures they expect from government. There was widespread agreement in the private sector on the importance of government planning and direction in the Australian computer industry, and the need for the development of a national computer policy. Underpinning the submissions was an acceptance of the notion that the Australian government ought strongly to support the electronics and computing industry as a key sector for future development. Understandably, of course, with submissions both from Australian companies and subsidiary companies of trans-national corporations operating in Australia, there was disagreement on issues such as local-content requirements in government tendering, and eligibility for forms of government assistance for firms who purchased goods manufactured by Australian owned companies.

A wide range of direct and indirect forms of government assistance other than bounties were commonly highlighted, especially R & D grants, fostering local venture-capital companies, selective taxation deductions, computer-education liaison, and local procurement preference policies designed to provide access to government markets. Despite the solidarity of views regarding future directions for the Australian computer industry, the IAC in its draft report ruled against the case for substantially higher assistance, though presiding commissioner, J. L. Sheaffer, submitted an impressively argued dissenting statement.[15] The final report extended the general line of argument against selective industry assistance. The Hawke government decided in July 1984 to replace existing import duties on computer products with a bounty, a considerable fillip for the domestic computer industry.

The Hawke Government commissioned a major private consultant to report on information technology. The Scott-Little report had as its prime terms of reference: to assess opportunity areas, formulate business strategies and make recommendations on which the government could implement its commitment to the development of sunrise industries by developing policies and actions to strengthen the position of Australia with regard to 'Information Technology.' According to this report, released in July 1984 (and thereby too late for 1984-85 Budget consideration), the estimated Australian information-technology related market of $3.2 billion in 1984 will have more than doubled in real terms by 1992, with software as the major growth area. The report further argued that Australia had so far failed to grasp opportunities in information technology, commonly referred to as IT, but whilst it acknowledged that 'the vast majority of IT products used in Australia are imported or are assembled here by foreign-owned companies' it tamely asserted that 'we should encourage and gain assistance from foreign owned companies' who 'should be encouraged to view Australia as a base for the export of particular products'. Rather than urging a strongly directed economically nationalistic plan, the Scott-Little report fell back on the failed rhetoric of the 1970s, and on the hope that technological transfer to Australia might work.

Scott-Little sensibly suggested that 'Australia should concentrate on those products and services which capitalise on our relative strengths'. The best development prospects were designated as special-purpose products, clearly differentiated products, integrated systems with high intellectual value-added content, and systems or products developed to meet the requirements of Australian conditions but which might have other regional applications. The advice tendered was that Australia should stay away from general-purpose computing and standard communication and that likely future growth areas were voice recognition and synthesis, flat panel displays, mass memories, artificial intelligence, parallel processing and image processing.

Australia, it was argued, could double its production and establish a local and export market worth $5.6 billion by 1992, as well as become the leading source of IT expertise for application in the developing countries of Asia and the Pacific in the 1990s. A five-point program to achieve these goals was detailed in the report – how to stimulate company growth, support innovation, support industry development through government procurement, develop a technology infra-structure and increase IT awareness. Scott-

Little claimed that with a $130 million government commitment to such a program, together with $40 million from industry, local-production target increases as ambitious as $10.5 billion over a 10-year period would be achieved.

There is undoubtedly substantial potential in terms of the development of a local manufacturing base in the telecommunications equipment market. According to industry submissions to the Davidson inquiry, the total telecommunications equipment industry is a $600 million market in Australia.[18] There are 11 large manufacturers and over 20 smaller manufacturers involved in the production of telecommunications equipment in Australia. Nine of the 11 major manufacturers are overseas-owned: three companies, Standard Telephone and Cables (STC), L. M. Ericsson (LME) and Australian-owned Amalgamated Wireless (Australasia) Pty Ltd combined have more than 50% of the market share. The fact that trans-national companies have established such profitable Australian subsidiaries demonstrates their view of the viability of the Australian domestic market. Government protection policy in this area rests on imported terminal equipment being subjected to a general tariff of 30 per cent and some network equipment attracts a 35 per cent tariff. A curious quirk of policy is that countries which are deemed as 'developing', such as Hong Kong, Korea and Taiwan, can effectively import telecommunications equipment duty free.

In Australia's development of a sustainable base for local manufacture, assembly and product development in the telecommunications field, Telecom's role is vital, as it exerts the most significant influence of any organisation in Australia on the size and nature of the Australian telecommunications manufacturing industry. Telecom must initiate, or be directed towards, a staged buy-Australian-made policy. Considering the long-term local manufacturing vacuum in this area, it will obviously take time to foster indigenous production of even part of Telecom's requirements.

Former Telecom managing director, Jack Curtis, summarised Telecom's long acceptance of an offsets policy. He said that 'rather than seeking the establishment of an isolated local industry to develop, design and manufacture to service the relatively small market in Australia, the policy has been to call worldwide tenders for major systems but to require a high component of local manufacture'.[19] The notion of an offset policy to develop local manufacturing must be scrapped in the long term. The Davidson report also pointed out that during the Fraser administration local preference applied to government purchases until October 1981, when under revised policies preference to Australian goods applied only to defence, strategic and high-technology industries. However, according to Davidson 'the high technology classification has not been applied to Telecom's purchases'.[20]

The market prospects in this field are substantial; Telecom, as national telecommunications authority, sustains sales in excess of $600 million annually to the private electronics manufacturing industry. Two telephone manufacturers in Australia, AWA and STC, account for 95 per cent of local telephone instrument-production capacity – in excess of one million instruments annually. Each year about 800,000 telephones are manufactured for Australia.[21] A production run of 400,000 telephones is regarded as cost-effective. Surely Australia could at least manufacture all of its own telephones. Why not work towards a competitive tender, for Australian owned manufacturing components only, in this sizeable and clearly viable local market? Telecom's purchase of PABX systems is from three major suppliers, Siemens, Philips and NEC, which have local manufacturing facilities but are internationally owned companies. A ministerial decision in 1980, during the Fraser period of toying with deregulation, prohibited Telecom from the facsimile and telephone-answering markets. The result was simply that all equipment in this area is imported. The role that Telecom must play in the future, especially in the context of the burgeoning telematics markets, is to undertake a pump-priming role by the allocation of an increased proportion of its tendering to indigenous companies.

Areas of manufactured equipment for satellite technology should also be designated as part of Australia's information-technology infra-structure. Australia has long been involved with international satellite technology. As a founding member of Intelsat, the international space-communications organisation, Australia had two earth stations operating by the end of the 1960s. Television companies and the Overseas Telecommunications Commission, responsible for international phone calls, have steadily used more of the Intelsat system in recent years. With the introduction of a domestic satellite system in 1985, Australia's local market for satellite equipment is one of the largest in the world. Yet the space components of the domestic satellite system, and related ground-terminal equipment, have predominantly been purchased from overseas companies. Whilst no one would naively suggest that Australia ought to set up in competition with America's NASA in space-shuttle launching, opportunities have been missed for fostering local satellite-equipment manufacture. The leading American satellite corporation, Hughes Communications Ltd, won the tender to construct three satellites for Australia. Mitsubishi won the Aussat tender for the major city earth stations, shared the Aussat tender for minor ground stations with Nippon Electric Co. and Codan, and Telecom preferred the Nippon Electric Co. for their ground stations. So, as McCracken has pointed out, in an area where Australia has major scientific strengths and significant local needs, the long-term failure to nurture local industry has led to virtually nothing happening in terms of local manufacturing development in satellite technology.[22] Aussat staff have argued that about $100 million has been allocated for contracts with Australian industry. Whilst such investments are, of course, productive in that they create jobs, the central point is that the bulk of this 'Australian component' goes to subsidiaries of trans-national corporations. While about 50% of the major city earth station contract with Mitsubishi is undertaken at their new plant in Ryde, NSW – local investment, local jobs – and STC locally manufactured wiring components for the satellites, these contracts did not go to Australian owned and controlled companies.[23] The twenty Aussat staff who worked in Los Angeles with Hughes Communications Pty Ltd on the construction and testing of the Australian satellites have been cited as an example of 'effective technology transfer to this country.'[24] However, if Australia fails to construct its own technological infra-structure within the next few years, this small pool of skilled talent will be limited in where their skills can be effectively applied locally, and may have to find employment overseas.

The Australian government decided in 1979 to move into the domestic satellite age. The Aussat project is an immensely expensive system, yet there has been bi-partisan neglect of exploring local manufacturing opportunities in domestic satellite technology for Australian owned and controlled manufacturing companies. Satellite technology is still in its infancy in Australia – it is not too late to develop policies to support local ventures, by local companies to maximise opportunities for Australians.

High Technology and the First Hawke Government

The first Hawke government offered a refreshing approach to science and technology in its early statements. Prime Minister Hawke's rhetoric in his official opening speech to the National Technology Conference in September 1983 indicated new directions in Australian government thinking about new technologies. He argued that 'sustained economic growth is going to require the progressive adoption of productive new technologies' and that 'the effectiveness with which we use new technology will determine whether we can reverse the long downward slide in our living standards.' Hawke proffered a sweeping commitment on behalf of the government, that 'the Labor Government works from the presumption that rapid technological change is desirable.' Policy decisions which emerged from the Department of Science and Technology, despite their refreshing sense of imagination and support for science-based industries, did not really go as far as the carte blanche commitment implicit in the Hawke rhetoric.

No one, of course, could reasonably expect a new government to build a high-technology base within its first two years, especially after decades of neglect by past governments. However, instead of designating information technology as a key component of the strategy to build a high-technology infra-structure, the first Hawke government pursued a moderateness of approach succinctly stated in the draft document, *National Technology Strategy*, that 'particular effort should be made to promote information technology.'[26]

What are the essential policy components of a government-planned strategy for high-technology development in Australia, with specific attention on information technology? One broad option would be substantially to increase total expenditure on R & D spending in the public sector, and to encourage greater research commitment in the private sector. In Australia, the total expenditure on R & D, from both the public and private sector, as a proportion of GDP, is well below that of the leading developed countries. OECD data shows that during the 1980s Australia's per capita spending devoted to R & D, a mere 1% of GDP, lagged well behind almost every major trading nation.[27] General Motors, Ford or IBM *each* invest more on R & D in the USA than the Australian government and industry combined. And, of course, as a product of its essentially internationally owned character, Australian business enterprises account for only 20 per cent of the national R & D effort: the OECD average is a 40 per cent commitment from the private sector to national R & D.[28]

Despite record national government budget deficits, a substantial increase in R & D spending is central to a strategy for economic reconstruction. The targets suggested in the draft discussion paper of *National Technology Strategy* – 1.5 per cent of GNP spent on R & D by 1990 and 2 per cent by 1995 – are too moderate. A substantial increase in R & D funds ought not be interpreted as a blanket policy towards the hopeful creation of a range of new and interesting sunrise industries, with vague marketing possibilities, nor a policy of let's hope we can find an Australian Edison in the 1980s. These funds must be an integral part of a strategy which attempts to build a more economically independent manufacturing and technological base for Australia.

In its first year in office the Hawke government allocated $16.7 million in project grants to companies through its Australian Industrial Research and Development Incentive Schemes, of which computer-communications received $5.5 million, as a useful but moderate beginning on a dollar-for-dollar basis with innovative companies. Steps were also taken by the first Hawke government to ensure greater availability of venture capital for companies who wish to invest in high-technology companies. One of the factors for the loss overseas of those information-technology products invented in Australia has been the conservatism of conventional financial sources. In the USA and Europe, venture or risk capital has been an important component for the success of new high-technology companies. Australia, until recently, had virtually no venture-capital market in high technology. We can only guess how many of those failed small computer software companies might have made their mark had they been given a longer financial lease of life through venture-capital support.

A committee of the Academy of Technological Sciences, headed by Sir Frank Espie, proposed a plan in 1983 to launch a private-sector venture industry to fund entrepreneurial schemes. In May 1984 the Hawke Government, in accepting the Espie plan, licensed seven Management and Investment Companies (MICs) with the right to raise tax-deductible investments in innovative Australian technology ventures. MICs exist under the auspices of the MIC Licensing Board. The first round of licence applications demonstrated a strong field of contenders, with 7 out of 32 successful applicants, at an estimated cost to government of $50 million.[31]

Investors in an MIC are allowed a 100% tax deduction on their investment in the hope that they can pick a high-technology company winner. There are some similarities in this scheme to the way in which the Australian film industry has been supported, in that the technology investor gambles on the success of the final product from seeing a prototype (the equivalent of a film script and prospectus). The new venture-capital market will encourage the diversification of funding from traditional sources, such as inherently conservative banks and insurance companies, towards merchant banks, accounting firms, solicitors, stock brokers and other entrepreneurs. This tax-deductibility strategy for private-enterprise speculators was one of the major reasons why Australia has been able to develop an indigenous film industry. MIC high-technology licensees, like Australian film-production houses, should only be indigenous Australian companies.

Government purchasing policy must also act as a catalyst to stimulate local production of high-technology products. Preferential purchasing from local suppliers by government departments is established practice in most countries, but not in Australia. It is ludicrous that Australia has such loose preferential policies with respect to government purchases of computer equipment, and that statutory corporations like Telecom and Aussat have only limited pressure on them to encourage indigenous manufacturing. Government tendering practices have unfortunately generally discriminated against local manufacturers in the way tender specifications are written and in the time allowed for tendering. Where short lead times are involved with tendering, overseas suppliers are better able to call upon a large range of products at short notice.

In the computer field, there is a common view that a 'buy Australian Act' should be introduced for the public sector. In submissions to the IAC, the Australian Computer Equipment Manufacturers' Association proposed the mandatory requirement that Australian companies be offered the opportunity to 'tender for government orders and be allowed to renegotiate the tenders if they are first rejected because of price' and that 'local manufacturers be given the right to seek a renegotiation of a government tender if it is apparent that no preference has been accorded comparable local products.'[32] While their first suggestion may be somewhat unrealistic, policies which take into account the national ownership of computer companies are desirable. The Australian Computer Society (ACS) has proposed that preference in hardware purchasing, for Commonwealth and State governments, be extended to a preference level of 50% to computer systems manufactured in Australia. In software, ACS proposed that access to government markets be used to stimulate local software-industry development, that government departments be encouraged to standardise their software, and that specific measures be used to enhance the offset program.[33]

The uncoordinated nature of government's forward computer-purchasing requirements was another criticism. Commonwealth purchasing-preference initiatives announced in December 1983, to operate initially on a trial basis in the computer field, stated that government buyers will be directed to examine the scope for purchasing locally developed equipment, that purchases will be planned well in advance and that the industry will be assisted to gain a better understanding of government funding procedures and requirements. In June 1984, the Minister for Administrative Services, John Brown, released a $250 million shopping list from Commonwealth Government departments and authorities for the next three years. This encouraging move needs to be strengthened by a stronger 'buy Australian' policy directive to all government departments throughout Australia.

To date, few local equipment makers, with the notable exception of AWA, have been successful in winning contracts in information-technology equipment with Telecom. In May 1984 Telecom, politically more astute than often given credit for, announced its intention to increase its support of communications-technology equipment designed and manufactured in Australia. Chairman Brack pointed out that about 90 per cent of Telecom's equipment purchases worth $1000 million a year were manufactured in Australia, but only a small amount of design work was undertaken locally since most of the companies involved were

Australian subsidiaries of overseas manufacturers. It was 'up to Telecom to create the opportunities' to redress the growing imports imbalance with greater Australian involvement in design and development, Brack announced.

Telecom must be held to these statements of intent. A month later, Telecom announced that the UK-based GEC group would be the major supplier for Australia's planned $6 million national videotex network, based on British Telecom's well established Prestel system. Had Brack's local encouragement policy been operational years ago, strategic planning for Telecom's videotex system could have incorporated a major manufacturing component for Australian companies.

Early in 1984, the Hawke government gave Telecom a qualified opening to embark on joint ventures with private enterprise. Under amendments to the Telecom Act, the Commission was 'given the power to form subsidiary companies that are involved in the business of telecommunications by the use of space satellites.'[36] The notion of encouraging public corporations to participate with private enterprise in developing new Australian industries is long overdue. Telecom ought to maximise its technical and marketing expertise in areas where it has strengths, including microelectronics, exchange equipment, advanced telephone equipment and data-processing equipment. Whilst it would be wrong to expect Telecom itself to become a manufacturing organisation, it ought to have a vehicle to convert its own research and development projects, which are funded by the Australian community, into successful local manufacturing ventures.

If the way of the future in information technology in Australia involves satellite equipment, pay-television decoders, videotex equipment, electronic funds transfer and electronic mail equipment, and the usual high demand for telephone terminal manufactured goods, how much of this manufactured equipment will be truly Australian? Do we simply accept that we are not competitive with overseas suppliers, or that better products are always cheaper overseas? We cannot embrace more new technologies in Australia, yet passively allow the process of manufacturing colonisation to continue. Though the rhetoric of the first Hawke government generally embraced information technology, and Barry Jones emerged as one of the nation's leading visionaries, he failed to get his technology strategy adopted as a centrepiece of government economic-reconstruction strategy. It was, of course, a government which inherited a record deficit, and was thus forced to agonise over competing ministerial claims in annual budget deliberations. Barry Jones failed to get full cabinet commitment for science and technology policy innovations on a major scale. But at least three worthwhile technology policy innovations must be acknowledged – the establishment of MICs to provide venture capital, the changed Australian Industry Development Corporation, which became able to take up equity in high-tech companies, and the increased Industry Research and Development grants. In terms of major capital outlays, however, the Department of Science and Technology fared badly. In the 1983-84 budget, total government outlays increased by 15.8 per cent but the Department of Science and Technology allocation increased by a mere 4 per cent and in the 1984-85 budget the scoreboard was: government outlays up 13 per cent, science and technology up 1.6 per cent. Not only was this area rejected in the budget context as a catalyst for economic change, but existing science programs and organisations failed to even hold their own. By comparison, on a percentage basis, the Hawke government in its first two budgets increased the funding to sport by 114 per cent.

There were several reasons for the failure of science and technology policies to be accepted as worthy of higher priority by the first Hawke government. Jones attempted to deflect criticism by telling scientists that they were 'collectively the wimpiest collection of lobbyists', who were inevitably out-manoeuvred in the political process by people pushing short-term political interests over long-term technological issues.

In the area of industrial restructuring and high-technology development, vexed policy issues cut across at least three government departments: the Department of Industry and Commerce, the Department of Science and Technology, and the Department of Communications. The Ministers who held the first two portfolios, Senator John Button and Barry Jones respectively, had philosophical and strategically different views about the nature and rate of technological change.

Barry Jones, as Minister for Science and Technology, underwent a change in his first year of office from the liberal pessimist of his book *Sleepers Wake!* to being an ambitious political technocrat.[37] As Minister, the Jones position became that 'the technological base of a nation *is* the economy itself' (his emphasis).[38] John Button's approach was not 'high technology at all costs', as in the revised Barry Jones position. Button argued that European experience had shown clearly that a policy solely directed at sunrise industries was futile. Button's broad view was that areas in which Australians would do well were those not subject to real competition (like tourism) and those industries where Australia has a lot of natural advantages in terms of resources. Though Button was supportive of the Jones position in public, and commendatory of the high profile Jones achieved for his portfolio, he did not go as far as Jones in terms of the desired level of government involvement in economic reconstruction. Rather than embark on a policy of 'make technology policy on the run', as Jones had advocated, Button's approach was more conservative and less all-embracing than the Jones position.

Button and Jones also deviated on fundamental notions about ownership and desirable power structures for Australian industry. Button had placed particular emphasis on 'introducing the most up to date and efficient technologies into existing industries.'[39] Button, for instance, before announcing his long considered master plan for the future of the Australian car industry, consulted with several major Japanese car manufacturers involved with the industry in Australia. Significantly, soon afterwards, Toyota, Japan's biggest car manufacturer, announced a $250 million investment over the next five years to stream-line the production of its Australian made cars. If Toyota's new plant in Australia in future employs few Australians, but many robots, John Button's argument will be that this sunset industry failed to compete.

By contrast, the Jones plan for reconstruction based on technological change was highly nationalistic. Barry Jones has consistently attacked the fact that so many technology-based industries in Australia are controlled by trans-national corporations and pointed out how this has inhibited the development of a dynamic high-technology infra-structure.

Button's position in the Hawke government, as Leader of the Government in the Senate, and as a member of the inner Cabinet, was politically stronger than Jones's. The Science and Technology portfolio had long been regarded as a junior ministry in Australian governments, though Jones made more impact on policy than any of his predecessors. Michael Duffy, as Minister for Communications, with ministerial responsibility for Telecom and Aussat, inherited a series of half-baked policy commitments related to new communications technologies and was essentially reactive with no apparent philosophical position. The inherent conservatism of departments like Treasury, Prime Minister and Cabinet and Finance frustrated all three ministers, but most of all Jones, who attacked what he called their economic agnosticism in supporting a bold new technological strategy. The first Hawke cabinet regarded Jones as a high flier who was good for the Government, but they were not prepared to accept his bold initiatives and the associated costs that came with what they essentially regarded as a hazy area of futuristic policy.

An overall assessment of the first Hawke government in the field of science and technology, ought to credit some promising innovations but achievement failed to match the rhetoric. The ALP major pre-election statement, the *National Recovery and Reconstruction Plan* showed a refreshing grasp of understanding traditional problems, and held the prospect of fundamental changes in direction. This plan declared that 'Australia must assert its own technological sovereignty' and that the 'ALP declares that the introduction, ownership and control of high technology should not be left to market forces alone.' At last, it seemed, an Australian Labor government might just develop policies for technological innovation and social worth based on an economically nationalist approach.

Jones, in his opening address on 'Technology and Australia's Future' to the Canberra Conference, told delegates that 'we have come to take it for granted that the commanding heights of our economy are in foreign hands, that our manufacturing capacity is largely for assembly only: that we drive cars, swallow drugs, and use computers designed by people who are cleverer than we are.'[40] Are we now resigned to multiple forms of technological subservience? This must surely be one of the crucial litmus tests for the second Hawke Administration.

5 The Media of Tradition: New Strategies for Old Players

Australia's pattern of commercial media ownership and control contrasts with its foreign-owned information-technology based industries in that commercial media developed around a foundation of a few Australian owned companies. Media is big business, and a few conglomerates dominate the traditional major media of press, radio and television. These groups are strategically well placed to diversify further into electronic information publishing. Historically, media conglomerates have tended to emerge out of successful family-owned newspaper businesses. The tendency towards press oligopoly has increased from 1903, when 21 capital-city daily newspapers in Australia were owned by 17 independent owners, to the mid 1980s, when 18 dailies are in the hands of only three owners.[1] Moreover, when radio licences became available in the 1920-30s, it was largely the successful newspaper companies who acquired them, and this pattern was perpetuated in the 1950-60s when predominantly the same groups were successful in also gaining metropolitan television licences. The end product is that Australia has the highest level of concentration of media ownership and control in the democratic world.

Notions of control and power in Australia's media institutions involve a different series of issues from those canvassed earlier about information technology. There is obviously a world of difference in discussing the role of a company which manufactures and markets computer hardware, to that of an organisation which has control of a potentially influential television licence. Finite spectrum space led to a licensing system in broadcasting, which assured that decisions about who would have access to the limited broadcasting frequencies was the province of the government of the day, rather than being subjected merely to the forces of the marketplace. Economic protectionism with favoured treatment by government of specified groups in broadcasting later emerged to complement the technical rationale for the licensing process in electronic media. However, broadcasting-industry legislation, unlike information technology, has long sought to exclude non-citizens from controlling Australia's electronic media.

A test case in the 1950s established a firm precedent to prevent foreign control of Australian broadcasting. In 1951, the Menzies government made foreign control of Australian media a major issue when the publishers of the English *Daily Mirror* and *Sunday Pictorial* sought the acquisition of a controlling interest in the Macquarie network of radio stations. A resolution of each Australian House of Parliament asserted that 'it is undesirable that any person not an Australian should have any substantial measure of ownership or control over any Australian commercial broadcasting station, whether such ownership or control be exercisable directly or indirectly.'[2] This principle has remained enshrined in broadcasting legislation; Section 90G (TVS 92D) of the *Broadcasting and Television Act* restricts any one foreign person to 15 per cent of the interests of a licensee and also restricts total foreign interests to 20 per cent. This has had the desirable effect of virtually keeping foreign interests completely out of Australian broadcasting.

Media Ownership and Control: Diversity in a Democracy?

Australia developed different patterns of ownership and control in the telecommunications and broadcasting industries. In telecommunications, the European model of common-carrier public institutions was adopted; three public institutions with different functions now share the national telecommunications statutory responsibilities: Telecom, the Overseas Telecommunications Commission (OTC) and Aussat Pty Ltd, the new satellite authority. Their roles are discussed in a later chapter. In broadcasting, two paradigms emerged in Western democracies – the private competitive market-based operation, of which the USA is the best example, as opposed to a model where a publicly owned broadcasting corporation is pre-eminent, such as in the UK, with the British Broadcasting Corporation (BBC). Australia developed a dual broadcasting system – a small group of private-enterprise media operators as well as a publicly funded national broadcasting institution, the Australian Broadcasting Commission (now Corporation) founded in the 1930s. In the 1970s, the ABC's monopoly on public-service broadcasting was banished with the introduction of a second public-service broadcasting institution, the Special Broadcasting Service, for ethnic or multi-cultural broadcasting, and also with the introduction of a new sector of public or community radio stations. The chart, Outline of Australian Broadcasting and Telecommunications on page 72, shows the schematic representation of the institutions, and their broad lines of legal and political accountability.[3]

The winds of change have made their mark on every Australian public telecommunications and broadcasting institution in recent years. Every institution on the chart has been involved in some way with major public inquiries in the past decade or so. This has not only been in the form of inquiries into the possible introduction of new communications technologies such as cable or pay television and the domestic-satellite system, but three public institutions – Telecom, the ABC and the SBS – have themselves been subjected to major institutional investigations: the Davidson, Dix and Connor inquiries respectively.[4] These inquiries are symptomatic of a contemporary crisis of confidence in the public models of communications institutions. The public-service telecommunications institutions, conceived to guarantee social equity of telecommunications services, and the public-service broadcasting organisations, with lofty aspirations related to public worth, are facing a process of fundamental reassessment.

The drive towards a more aggressively competitive marketplace paradigm, which underpins these pressures for change, has derived its renewed contemporary strength from two major sources. First, the general preference towards economic and ideological assumptions which elevate the private market-based system over that of public institutions. Second, the introduction of new technologies of broadcast spectrum propagation are being used to challenge a major assumption which contributed towards the past legitimisation of public-service broadcasting – the finite space of broadcasting frequencies.

There is now a line of argument, albeit unconvincing, that technologies such as cable television and new satellite-delivered

CHART 5:1 OUTLINE OF AUSTRALIAN BROADCASTING AND TELECOMMUNICATIONS

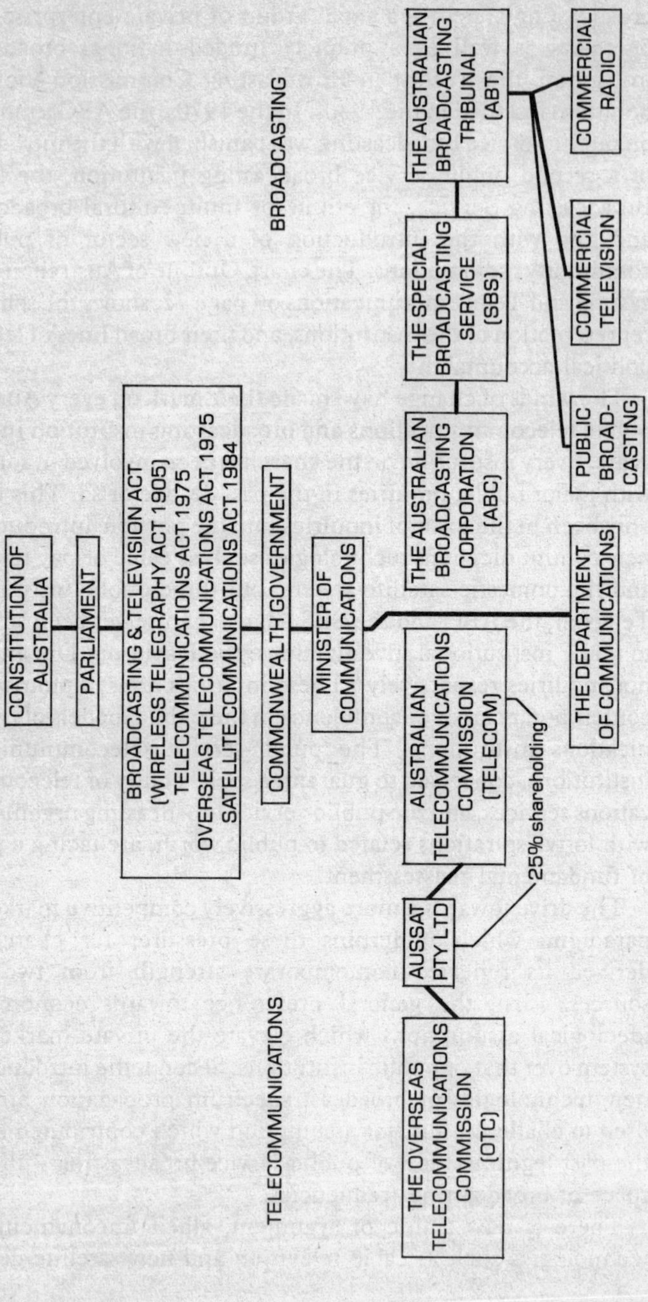

television services will open up such a spectacular array of new broadcasting services, that audiences can be liberated from patrician public broadcasting. According to private-sector protagonists for change, the technological revolution can offer an unprecedented cornucopia of electronic services with a wider range of electronic gratifications in a new era of abundance. Before these arguments can be properly pursued, we need to look first at the private-sector players in the media marketplace, examine notions of ownership and control, and then examine the present bases of media power.

Australia's commercial media is dominated by four large corporations: Consolidated Press Holdings Ltd, The Herald and Weekly Times Ltd, John Fairfax Ltd and the News Corporation Ltd. These four largest media groups in Australia have dominated the commercial ownership and control of traditional print and other broadcast media, but they also have extensive involvement in music, video, book publishing and several new information services such as videotex and teletext. Corporate strategies have varied regarding their overseas media extensions, and in Australia their non-media diversifications have included property investments, cattle ranching, coal and oil companies, travel and transport industries.

These four major media companies own Australia's major newspapers, both metropolitan and suburban, most magazine publishing and the major newsagency, Australian Associated Press. Three of the four corporations own all of Australia's 18 metropolitan daily newspapers (Consolidated Press is not involved), and the oligopoly of four owns 10 of the 14 Sunday and major weekly papers, together with extensive interests in regional and suburban papers. Australia's chief newsprint supplier, Australian Newsprint Mills, is jointly owned by the Herald and Fairfax groups. The same four large media groups effectively control all of the commercial television stations in Sydney and Melbourne, and one station in Brisbane and Adelaide. The three metropolitan networks in Sydney and Melbourne (7, 9 and 10) occupy an extraordinarily dominant market position in Australia's commercial television system. For example, these six television channels have a combined market potential of 42 per cent of the total national audience. In 1982-83 the program expenditure of these six stations was 62 per cent of total Australian commercial television station revenue and 55 per cent of the national aggregate commercial-television advertising revenue.[5] In radio, of the metropolitan radio stations, the same four groups control 11 out of the 37 metropolitan AM radio stations and they also operate nine regional radio stations.[6]

Old arguments about the effects of excessive levels of concentration of media ownership and control have been revived in the context of the new technologies. It can be convincingly demonstrated that Australia has abnormally concentrated media ownership, but it is a more difficult proposition to justify that this is a matter of real concern. The arguments against high levels of concentrated media ownership and control involve broad political, cultural and ethical judgements and they rest heavily on perceived or potential dangers to the community. They tend to be vehemently rejected by members of media management on the grounds of canvassing the possible rather than the actual. Wherever the debate about concentrated media ownership is couched in narrow legal terms, protagonists for change are seen as being out of touch with the real world, or as radicals who cannot deliver tangible evidence to conclusively prove their assertions. Yet, as Lord Reid, the famous English judge, once said in defence of the principle of natural justice, there is a 'perennial fallacy that because something cannot be cut and dried or nicely weighed or measured therefore it does not exist.'[7]

Arguments against concentrated media ownership rest on the premise that the unchecked concentration of power is undesirable. Media organisations are controlled by a small group of people, unelected and unaccountable in many ways to the community at large. On the assumption that the media are influential in people's lives, it is the arbitrary power, and the potential for its abuse, which is of central concern.

A second line of objection concerns barriers to entry of newcomers. Henry Mayer has argued that with newspapers, the major groups have pooled their news sources, shared ownership participation of newsagencies and newsprint mills, and generally exercised such tight de facto control over distribution outlets that this has made the possibility of competition harder for the entry of outsiders.[8] With radio and television, the scarcity of broadcasting licences has reinforced the entrenched market position of members of the oligopoly.

A third area of concern is that the present system lacks full diversity of expression in media outlets. What the community sees as issues depends in part on who constructs agendas in the media, and how this process of construction occurs. The Australian media system is essentially dominated by a commercial ideology. That is not to say that there are not significant differences of opinion within commercial media, and also within one commercial media organisation, but the chances of wide diversity are greater where there are a significant number of different players involved, with different perceptions of the world. Commercial media institutions generate internal pressures which are a product of the character of their ownership. It is not that a few bosses direct editors and journalists about what they may or may not publish or write, but that media personnel internalise the values of the organisation and tend to become conditioned by their occupational environment into conventions of commercial uniformity.

We read a lot about 'freedom of the press' in the press itself. Yet this ought to involve more than attacks on unwarranted forms of political intervention or calls for liberalised laws of libel and defamation. The notion of freedom of expression ought to involve the broadest range of interpretations possible, including the presentation of different values from those implicit in commercial media, and participation by groups who have been deprived of access to the media. If freedom of expression is cherished in a democracy, then a media system must guarantee ideological diversity of expression. New technologies in electronic media create opportunitites to open up the system, but if the newcomers are ideologically similar to the old players, no diversity is assured. This point is often missed by those who propose that the new technologies of abundance will solve old problems of concentrated media ownership, control and influence. It all depends on who the new players might be. The danger is that they will merely offer more of the same. Ideally, a communication system should guarantee an ideological mix of media owners and controllers, with individuals and groups who have the right of access and who have full opportunity to express a diversity of political, social and philosophical stances.

The issues of media monopolies were canvassed in a State government inquiry into newspaper ownership in Victoria established by the Hamer Liberal Government in 1981. This inquiry arose in the aftermath of the News Corporation's attempted takeover of the Herald and Weekly Times Group in 1979. This bid failed, but the incestuous share trading between newspaper corporations left Victoria with further increased concentration of newspaper ownership. Fairfax then owned about 58 per cent of David Syme Ltd, publishers of the *Age,* but to protect the Herald group from takeover, the Fairfax group bought 14 per cent of the Herald group, which in turn owned 14 per cent of David Syme Ltd. The Fairfax acquisition had thereby further intensified the ownership of Melbourne's daily newspapers in a way which potentially threatened the independence of the *Age.* Phillip Adams convened a group called 'Friends of the *Age',* arranged a packed Town Hall of *Age* staff and protesters, labelled Melbourne a newspaper-company town, and called on Fairfax to sell its shares in the *Age.* Retired judge Mr (now Sir) John Norris was invited by the State government to investigate the ramifications of ownership changes in Victorian newspapers, and to advise on whether any form of regulation or restrictions were considered warranted or desirable.[9]

Predictably, the major newspaper groups formed a united bloc, opposed to any proposed form of regulation of newspaper ownership on the grounds that no deleterious effects of concentration of ownership could be established. Ranald Macdonald, then managing director of the *Age,* emerged as the black sheep of newspaper management to argue in a personal submission that the Victorian Government ought to set up a statutory authority to examine mergers, acquisitions and amalgamations in Victorian metropolitan newspapers. Norris not only accepted this recommendation, but extended it to include all newspapers published in Victoria with a circulation of over 100,000 copies. He premised this recommended course of action on two central arguments adopted by the 1961-62 British Press Royal Commission – that the major dangers associated with concentration of ownership and control of the press are the loss of diversity of expression, and 'the power of a very few men to influence the outlook and opinions of large numbers of people.' The Norris report recommended the establishment of an independent Press Amalgamations Authority to scrutinise future share transactions which involved the ownership of newspapers by the existing corporations. This would have provided supervision of market structures, but not government involvement in the editorial affairs of newspapers. Despite the thoroughness of the Norris Report, and its moderate recommendations, successive Victorian State Governments, both Liberal and Labor, have ignored it.

In broadcasting, the policy history shows rhetorical political commitment and legislative intent to restrict the levels of concentration of ownership and control, but abysmal failure in practice. Such moves can be traced as far back as 1935 when, soon after the introduction of commercial radio, the Lyons Government introduced limitations on the number of radio licences that any one person could hold.[10] Similarly, in the formative days of Australian television, the 1953 Royal Commission acknowledged the need for limitations on the concentration of ownership and control of commercial television stations.[11] Legislation in 1960 and 1965 limited the total number of radio and television channels which could be owned by one interest. In the mid 1970s the Green Report, which led to a restructuring of the regulatory framework of Australian broadcasting, and the creation of the Australian Broadcasting Tribunal, recognised the desirability of 'avoidance of undue concentration of interest'.[12]

The difference ought to be made clear between the terms, ownership of a newspaper, and control of a radio or television channel. The Commonwealth government chose to license specific broadcasting frequencies to select groups for reasons of limited spectrum space and economic protectionism. Members of the oligopoly have established subsidiary companies which can control two television licences and these are generally automatically renewed by the Australian Broadcasting Tribunal on a three-year basis. In the early days of television-licence allocations, the Australian government clearly had no realisation that television would become such an influential and profitable medium. The Government could have, but did not, use licensing as a mechanism to ensure a diversity of television controllers. Control can change hands by the purchase of the company which controls a television licence. Share-market operations have allowed the wealthy players to acquire the most attractive licences; for example the Murdoch group in 1979-80 sold its Adelaide and Wollongong assets to acquire companies in Sydney and Melbourne which gave it strategic control of the Channel Ten network. Despite the political rhetoric about restrictions on the concentration of media ownership and control in party platforms, the share-market horse trading for television licences in recent years has taken place in such a way that the groups have been essentially unaccountable for their acquisitions. The history of the licensing of Australian commercial television shows that progressively a few corporations have moved to consolidate their privileged access to public assets.

The legal approach to restrict the levels of concentration of ownership and control in electronic media has been to essentially place upper limits on the number of licences held by an individual or group. Legislation has attempted to prevent excesses of horizontal integration; for example, one company may only control a maximum of four commercial radio licences in any one state and a maximum of eight throughout Australia.[13] Yet the issue of cross-media concentration, where one company controls substantial press, radio and television interests, has been virtually ignored by legislators and policy makers.

The history of the allocation of commercial television licences in Australia shows decisions which accentuated the degree of cross-media ownership and control. Briefly, Section 91 of *The Broadcasting and Television Act* 1956, under which commercial television was established, introduced a numerical limitation of control of no more than one television station in any capital city or two stations in Australia. In the initial grant of licences in 1955 there were eight applicants for the Sydney licences and four for the Melbourne licences. Two licences in both Sydney and Melbourne were granted: ATN-7 Sydney to Amalgamated Television Services Pty Ltd, TCN-9 Sydney to Television Corporation Ltd, HSV-7 to Herald-Sun TV Pty Ltd and GTV-9 Melbourne to General Television Corporation Pty Ltd. In the beginning, then, no applicant was granted more than one of the licences available.

When the licence applications for Brisbane and Adelaide were reviewed in 1957, the Australian Broadcasting Control Board recommended to Government that only one licence be granted in each city, and to a company 'able to demonstrate that it is substantially locally owned and not controlled in any way by any of the companies holding the licences for existing stations or significant shareholders in these companies'.[14] The Menzies Government ignored the ABCB recommendation to call for fresh applicants in the hope that local companies would then apply (even though local control of television was stated government policy) and decreed that two licences should be granted in each city, and that they should be selected from existing applicants.[15] This benchmark decision set the destiny for control of commercial television, in that the successful Brisbane and Adelaide licensees were predominantly newspaper-based companies who then came to hold interests in more than one television channel.

Between 1960 and 1965 additional licences were granted in Sydney, Melbourne, Brisbane, Adelaide and Perth, which further accentuated the levels of concentration of control in commercial television. Amendments to the *Broadcasting and Television Act* 1960, aimed at tighter definition of the term, 'control', actually led to share transactions which resulted in a number of companies acquiring interests in more than two licences. The 1960 amendments also restricted the Minister's authority to refuse approval of a change in shareholding. The ludicrous situation of ministerial licence allocations became clear with the later grant of a Brisbane television licence. On the recommendation of the Australian Broadcasting Control Board, Ansett Transport Industries (ATI) was an unsuccessful applicant, but before the Minister could grant the licence as recommended, ATI had purchased a controlling interest in the company of the successful applicant. The Minister was powerless to act, and ATI thus controlled two metropolitan television licences. Further amendments to the Act in 1965 were introduced to render Section 92 more effective. The concept was introduced of a 'prescribed interest' at a level of five per cent of share interests or voting rights. Ministerial discretionary power to regulate shareholdings of licensee companies was re-established.

Ownership and control provisions of the Act have been essentially the same since 1965. Section 92 of the Act limits the number of 'prescribed interests' which a person may hold in commercial television licences to one licence in the capital city of a state, and to a total of no more than two licences. The cornerstone of policy to prevent further concentration of ownership and control, the two-station rule, ignored cross ownership and is also ludicrous on commercial grounds. It simply ignored the size of the market audience served by licensees. How, for instance, could a Toowoomba and Bendigo television licensee be comparable on the two-station rule to a licensee who controls a station in both Melbourne and Sydney?

The Australian Broadcasting Tribunal has also made several

judgements against companies with multiple broadcasting interests on the grounds of public interest. Such challenges to influential groups have resulted in amended legislation to curb its authority in specified areas. In what became known as 'the 2HD case' the ABT refused its consent to the transfer of a Newcastle commercial broadcasting licence on the grounds that the transfer would result in a degree of control that would be contrary to the interests of the Newcastle public.[16] On appeal the ABT decision was upheld in the High Court. Again in 1980 the ABT refused, on public-interest grounds, to approve the News Group acquisition of a 50 per cent interest in ATV-0 Melbourne. The ABT ruled that the common ownership of a Sydney and Melbourne television channel would have an adverse effect on the freedom of choice of other members of the network. The decision was overturned on appeal to the Administrative Appeals Tribunal, and News Group were then able to form the Sydney-Melbourne Channel Ten network. In 1981 the Minister for Communications, Ian Sinclair, introduced amendments to the Act which downgraded the significance of public interest in undue concentration of influence. The deplorable effect of the amendments was virtually to remove the ABT's discretionary powers on cross-media interests.

The Murdoch Channel Ten case brought into focus for the first time the issue of networking as a factor in the effects of concentration of television ownership and control. Networking is a business arrangement between a group of channels who co-operate for the acquisition of local and overseas programs, for program production and distribution, and for promotional purposes. The television stations in mainland capital cities operate within three networks (7, 9 and 10) where individual stations made a contribution towards program purchase based on the respective size of their markets. Networking provides the commercial advantage of economies of scale, and it must be acknowledged that individual small stations could not afford to maintain their present range of programs if networking arrangements did not exist. However, Australian regulatory broadcasting authorities have expressed their concern that networking could lead to undesirable business practices and particularly that the larger metropolitan stations could exercise undue influence over smaller local television operators.

These dangers have been renewed in the context of the introduction of Australia's domestic-satellite system, where the danger is that further power will be concentrated in the hands of the Sydney and Melbourne channels, thereby affecting the autonomy, commercial viability and local programming of regional stations. An extensive report dealing with these issues was handed down by the Australian Broadcasting Tribunal in July 1984.[17] This argued that the Sydney-Melbourne based networks had de facto control over all Australian commercial television. The networks contribute 60-80 per cent of the costs of acquiring Australian and overseas television programs, relegating other stations to a subsidiary decision-making role. The Tribunal pointed out that 'the metropolitan networks commission almost all entertainment programs produced in Australia; they control the sale of this material to the rest of the country's commercial channels; and they acquire a considerable proportion of overseas material which is broadcast on Australian commercial television.'[18] A corollary is the lack of opportunities for new television enterprises to enter the market.

The implications of networking have been largely ignored by politicians and policy makers, and networks have evolved outside the legal structures of Australian commercial television. Networking is unrecognised in the *Broadcasting and Television Act* and the Australian Broadcasting Tribunal has no power to regulate networks. The spirit of the two-station rule, to spread ownership and control among many groups, has been countermanded by operational business practices. The effect has been that, instead of having diversified control of commercial television, with a variety of autonomous licences in the hands of local interests, networking practices have contributed towards the development of a national commercial television system which was never envisaged. Moreover, the introduction of domestic satellites, which can technically provide the means by which three commercial television services can be made available to virtually all Australian residents, would create further imbalances of influence between the Sydney and Melbourne stations and the rest of Australia. The media monopolists are not only here to stay, but are strategically well placed to further consolidate their stranglehold.

Media Monopolies, Market Moves and New Technologies

Despite major takeover attempts initiated from both within the oligopoly and from without, the big four have remained firmly entrenched. There are examples of major investments by oligopolists to prevent takeovers, even of their major competitors, which might substantially alter the balance of power in the market-place: status quo, in terms of known competitors, provides a measure of corporate security. However, several major corporations which are presently without any media shareholding interests eagerly await openings which the new technologies might bring.

The Herald and Weekly Times Ltd is the grandfather octopus of Australia's commercial media. The Herald group is unusual for such a dominant newspaper-based enterprise in that it never had the strong founding family or owner equity participation so typical of successful newspaper conglomerates around the world. Despite its size and spread, the group has lacked a strong public profile and its long-term chairman and chief executive, Sir Keith Macpherson, never etched the indelible public image of a Rupert Murdoch or Kerry Packer. Under Sir Keith, the Herald group pursued a 'steady as she goes' approach, with a preference for the strengthening of existing corporate holdings rather than the pursuit of a more risky growth-oriented strategy. The group's financial performance over the past decade has been stable but stodgy, which has created some agitation about the rate of return on funds invested, especially from those shareholders who point to the spectacular growth in other media corporations.

Herald management has been forced in recent years into looking over its shoulder for corporate raiders. In November 1979, the Herald group successfully beat off a bold takeover bid from News Corporation, and again in December 1981, the Robert Holmes a Court Bell group failed in its bid to gain a 50.1 per cent control. Takeovers of the Herald group have been made more difficult by the intricate interlocking nature of the ownership of subsidiaries: the HWT owns 42 per cent of Queensland Press Ltd, which in turn owns 27 per cent of the capital of HWT, while the Advertiser Newspapers Ltd owns 9 per cent of HWT, which in turn owns 35 per cent of the Advertiser.[19] The holding costs of HWT's investments in other media companies has caused a substantial increase in interest costs and made the corporation one of the least attractive media stocks for investors. The HWT group provides a clear example of the strength of cross-media interests that a conglomerate can maintain in one city – Melbourne.[20]

TABLE 5:1 AUSTRALIAN TRADING STATEMENT: MAJOR MEDIA CORPORATIONS 1982-3 AND 1983-4

RANK	COMPANY	NET PROFITS $m		+/-%	REVENUE $m		+/-%
		1983/84	1982/83		1983/84	1982/83	
1	THE HERALD & WEEKLY TIMES (a)	32.7	24.4	+34.1	457.8	392.3	+16.7
2	JOHN FAIRFAX	30.7	15.3	+100.7	595.5	483.8	+23.1
3	NEWS CORPORATION (b)	18.8*	13.3*	+41	658.5	515.8	+27.7
4	CONSOLIDATED PRESS HOLDINGS	18.3*	14.5	+26.2	460.0*	425.6	+8.1

* estimate –

(a) financial year ends September 30.

(b) net profits is for Australian media interests only, revenue is for all Australian interests.

Sources: Annual Reports, *B & T Weekly*, and *The Journalist* March, 1985.

Newspapers - 100% of the daily *Sun News-Pictorial*.
- 14% of the proprietor of the *Age* (until April 1985).
- 100% of the only evening daily, the *Herald*.
- 50% of the *Sunday Press*.
- 100% of the proprietors of ten suburban newspapers.
- 100% of four other weekly publications.

Radio - 100% of the AM station, 3DB.
- 10% of the FM station, 3EON-FM.

Television - 85% of the licensee of the station HSV-7.

This list represents only its long-held prime commercial assets in Melbourne, but it has extensive media interests throughout Australia.

The overall profitability of the HWT group has been handicapped by its somewhat cumbersome business alliance with the Fairfax group in the Channel Seven network. Despite its initial strategic holdings in commercial television, the HWT group subsequently failed to lock up total control of a Sydney-Melbourne integrated television network in the way that both the Packer and Murdoch organisations have successfully achieved. The HWT group has an 85 per cent interest in Herald-Sun TV Pty Ltd, the licensee of HSV-7 in Melbourne, together with a 59 per cent interest in ADS-7 in Adelaide, following a successful takeover in March 1983 of Television Broadcasters Ltd.

The HWT group has long been the biggest publisher of daily newspapers throughout Australia. Two of its daily newspapers in Melbourne, the morning *Sun News-Pictorial* and the evening *Herald*, long enjoyed combined circulations in excess of one million daily, until 1981. However the *Herald*, a better than average evening newspaper, has in recent years suffered the sharpest circulation decline of any Australian newspaper. HWT group publications dominate the daily newspaper markets throughout the smaller states – Adelaide's only morning daily, the *Advertiser*, published by Herald-associate company Advertiser Newspapers Ltd, Brisbane's *Courier-Mail* and *Sunday Mail*, published by its Brisbane associate Queensland Press Ltd, Perth's *West Australian*, *Daily News* and *Weekend News*, and the Hobart *Mercury*, through its shareholding in Davies Brothers. Other Herald-group print activities include suburban newspaper interests with

Standard Newspapers Ltd, regional newspapers such as the morning dailies *Geelong Advertiser* and *Bendigo Advertiser*, a national magazine stable which includes *Australasian Post*, *Australian Home Beautiful*, *Your Garden* and *Prime Time*, and a range of Pacific Island weekly and monthly publications.

The HWT group is one of the largest shareholders in Australia's largest news agency, Australian Associated Press (AAP), the principal supplier of copy to Australia's media outlets. In 1983 AAP was split into two companies, AAP Information Services Pty Ltd, to maintain AAP's established operational services, and Australian Associated Press Pty Ltd, the investment arm of the agency. In January 1984 a new company, AAP-Reuter Communications was formed by AAP Information Services (52%) and Reuter (45%) to expand and develop electronic communications, including satellite links. The changes were made to facilitate the expansion and further diversification of AAP into broader and profitable services. Though often described as Australia's 'independent' national news agency, AAP's principal shareholders are the Herald and Fairfax group companies, as well as News Corporation Ltd. Under the former arrangement, members were charged for their news from AAP according to a formula based on state population and their number of titles held, together with a levy to make up the shortfall at the end of the year. Under the new arrangement, the principal shareholdings in AAP Pty Ltd remain as HWT group and the Fairfax group, each with a 42 per cent interest, and News Corporation with a 12 per cent interest; and in AAP Information Services the major shareholders are the HWT and Fairfax, 39.9 per cent each, and News Ltd, 15 per cent. The latest AAP company plans to operate a multi-directional microwave system (MDS) in the major capital cities to provide a data-distribution system developed by Reuters. It has bought exclusive rights in Australia for the dissemination of all Reuter financial and business services for the next ten years. AAP is also in the process of establishing an extensive network of earth stations and it will be the largest user of data delivered via Australia's domestic satellite. These kinds of moves show considerable awareness by oligopolists of the growth prospects of information as a commodity.

AAP developments are a collaborative venture by several media corporations, but the Herald group has played a large part in these initiatives. The HWT group is widely perceived as being a starchy organisation with a conservative management. The group displayed considerable sensitivity to criticism during the Norris inquiry hearings in 1981, with Sir Keith Macpherson spending three days in the witness box to defend the company's activities, advised by a bevy of top legal counsel. Other research and presentations to government inquiries on new communications technologies have been essentially left as a part-time job to former executive editor of the Melbourne *Herald*, John Morgan. His comment about the possible exclusion of existing groups from cable-television licences, at the Australian Broadcasting Tribunal national inquiry into cable television in 1982, may have been prophetic commentary on the future of the HWT group itself. Morgan said 'to lock us out could be a sentence of slow death. We have no wish to be left manufacturing gas mantles while the rest of the world is making electric light globes'.[21]

News Corporation Ltd seems to be regarded by stockbrokers as the world's most dynamic media property. With its 80 newspapers and magazines scattered across the USA, the UK and Australia giving a combined circulation of three billion annually, News Corporation Ltd bills itself as the seller of more periodicals than any other publishing company in the Western World. Its pioneer and Australia's most successful entrepreneur, Rupert Murdoch, has built a remarkable international media empire during the past thirty years. The diversified activities of News Corporation include newspaper and magazine publishing, broadcast television, printing and papermaking, newsprint warehousing and transport, airlines and diversified transport, oil and gas production and exploration.

Rupert, son of successful journalist Sir Keith Murdoch who built the HWT group into Australia's largest media corporation, began his newspaper climb with the Adelaide *News*. In 1960 he moved to Sydney and acquired the regional publisher Cumberland Newspapers, followed by Mirror Newspapers, publishers of the popular afternoon tabloid, the *Daily Mirror*. In 1964 Murdoch launched an ambitious national daily newspaper, the *Australian*, believed to have never made a profit in its twenty-year existence. In 1972 he extended his Sydney newspaper base with the purchase of the Sydney *Daily Telegraph* and *Sunday Telegraph* from Sir Frank Packer. In 1979 Murdoch decided to act on the long-felt disadvantage of not having a major television network to complement his chain of newspapers. He launched an audacious takeover bid for the HWT group, but was defeated by 'Friends of the Herald', an alliance of arch-rival John Fairfax & Sons Pty Ltd and Queensland Press. Subsequent to this failed bid, he turned his attention to Ansett Transport Industries in which his interest gave him control of Melbourne's ATV-0 and the Melbourne-Sydney television axis he wanted. More recent Australian expansion includes the launch of a new daily Brisbane newspaper, the *Daily Sun,* in 1982, and a $12 million one-third equity interest in Computer Power, a Melbourne-based computer company which was awarded the monopoly right in 1983 to produce Australia's first national legal retrieval data-base system.

Murdoch's international push began in the UK in 1969 when he bought the weekly *News of the World* and followed with acquisition of a London daily, the *Sun.* Both have increased their circulation and earning power under Murdoch management; the *Sun* became the largest selling English-language daily newspaper with daily sales exceeding 4.1 million (June 1983 audit) and *News of the World* with its four million circulation, the largest selling newspaper in the English-speaking world on Sunday. Murdoch rocked the old guard when England's premier thunderer, the *Times,* fell to him in 1981, but it is doubtful whether it will ever become profitable. In the USA, News Corporation began its acquisitions in 1973 with the *San Antonio Express* (Texas), a morning newspaper, and its afternoon companion the *News.* The launch of a weekly tabloid sold at supermarkets, the *Star*, followed, and in 1976 Murdoch bought America's fourth largest daily paper, the *New York Post.* The *Boston Herald* was also acquired in 1982 and the *Chicago Sun Times* in a US$90 million cash purchase late in 1983. There is a consistent formula in his approach. Buy a struggling paper, slash its expenses, replace staid news with sport, sensationalism and more 'readable' items, then embark on major advertising campaigns and add the News Corporation management style. Mike Royko, star columnist of the *Chicago Sun Times,* walked out soon after its takeover with the classic line that 'no self-respecting fish would be wrapped in one of Mr Murdoch's newspapers.'

Rupert Murdoch is an exceptional businessman and the foundation of this phenomenal growth is the 46 per cent equity of News Corporation through the wholly owned family private company Cruden Investments. In Murdoch's world of high finance, newspapers are far from dead – newspaper companies have contributed substantially to total group-revenue growth of $281 million in 1980 to $1,503 million in 1982-83, for a group-profit increase of $26.5 million in 1980 to $89 million in 1982-83. The Australian assets have provided a strong cash flow for financing the overseas expansion of News Group: of the total revenue of $1,503 million in 1982-83, $516 million was contributed from Australian operations, $648 million from the UK and $339 million from the USA. There is little doubt that it is Murdoch's intention to expand as much as possible in North America rather than the UK or Australia in future. The 1982-83 financial year saw record results for News Group publications in America – the *Star, New York Magazine, San Antonio Express, Houston Community Newspaper* and the *Village Voice.* There were also dramatic reductions in the losses of the *New York Post.*

One of the major corporate strategic decisions for News Corporation in the future is their diversification into electronic and information services. In 1983 in the UK, News Corporation acquired 65 per cent of Satellite Television (SATV), a small company which had somehow managed to acquire a transponder on a European Postal service satellite. The company then converted itself into an organisation which delivered English-language commercial television programs via satellite to cable systems in Europe. Murdoch moved in and bought a 65 per cent shareholding of this undercapitalised company in 1983 at a cost of $8 million, together with assurances to invest substantially in future satellite-delivered programming and marketing in Europe. This move assures the News group of a strategic place on the five new domestic satellites to be shared by seven European governments.

In the USA, the News group made bold preliminary moves into satellite-delivered television programming in 1983. Murdoch had decided to get an early foothold in one of the few neglected gaps in the US television market – those homes too remote to receive cable television but probably prepared to receive programs delivered via satellite to roof-top dishes. Originally, direct-broadcast satellites were due to begin in 1986; the FCC had authorised several companies, including the giant space corporation, Comsat, to operate new high-powered 200 watt satellites to provide small inexpensive dishes which consumers would need to pick up television signals. With typical Murdoch bravado, Skyband, the American broadcasting subsidiary of the News group, agreed to lease five satellite transponders from Satellite Business Systems at a cost of US$75 million over six years. The first down payment of US$1.07 million to SBS was made in October 1983, but then News retreated from the deal. The Skyband idea was that subscribers would lease a satellite dish three and a half feet in diameter and pay about US$25-30 a month for a programming package. The logistical problems which led to the retreat were reconsideration of the costs of installing and maintaining the dishes, and reservations about sufficient sources of programming. Skyband paid SBS US$12.7 million in compensation, and the whole deal cost News group an estimated loss of US$20 million.

The Skyband project was also behind Murdoch's failed attempt to take over Warner Communications in 1983. This bitter battle gave Murdoch a $40 million profit, but he failed to acquire the programming capacity that Warner's extensive film library would have provided. He later moved to become a half-owner of 20th Century Fox in a $355 million deal. Clearly the Murdoch organisation is gearing itself for further challenges in the electronic media field and for further expansion possibilities in the entertainment and knowledge industries. With Murdoch, it is continually a question of which move next on whom.

Kerry Packer's key management strategy of Consolidated Press Holdings Ltd has been to use the cash powerhouse of Australia's long-standing most successful television network, the Channel Nine network, to build a diversified commercial empire. His corporate planning has differed from Rupert Murdoch's in that Packer has essentially preferred to work towards expansion and diversification within Australia rather than planning an ambitious international push. Like Rupert, Kerry was the son of a newspaper tycoon. Sir Frank Packer, at the insistence of sons Kerry and Clyde, reluctantly agreed to sell the *Daily Telegraph* and *Sunday Telegraph* in 1972 to strengthen the group's holdings in television, which has provided bonanza profits. The Packer organisation has a broad array of media assets, not only in the traditional press, radio and television, but also in film processing, film production, record distribution, videotape recording, graphic arts, and media sports marketing. It was, of course, Kerry Packer who achieved international notoriety in the late 1970s when, frustrated in his attempts to secure Test cricket rights for Channel Nine television, he boldly contracted the world's best cricketers to the group's own World Series televised cricket.

Commercial television is a great investment for the entrepreneur: it generates cash rapidly, there is no need to finance stock, licences are scarce and granted only to the privileged. Consolidated Press Holdings Ltd holds all of the stock in Publishing and Broadcasting Ltd, which in turn holds all shares in the licensee company of TCN-9, Sydney. In Melbourne, GTV Holdings Pty Ltd, a wholly owned subsidiary of PBL, holds 80% of the shares in the licensee company of GTV-9, with Consolidated Press holding the remainder. Packer has moved out of daily newspapers, but owns the *Australian Women's Weekly* which consistently sold 850,000 copies weekly in its heyday and which has still maintained market leadership and solid profitability since its conversion to a monthly magazine. Other magazine interests include the *Bulletin, Australian Business, Mode, Belle, Australian House and Garden, Cleo, Modern Motor, Wheels,* and other specialist publications. Consolidated Press has kept its extensive interests in suburban and regional dailies in NSW and WA; in 1981 it acquired the balance of shares outstanding in Manly Daily Pty Ltd, publishers of the *Manly Daily,* probably Australia's most profitable suburban newspaper. The group's radio stations have also proved to be good cash-flow generators: Consolidated Press has a full shareholding in 3AK Melbourne, 6PM Perth, 6AM Horsham, 6KG Kalgoorlie and a 53 per cent interest (31 December 1983) in 6GE Geraldton.

In the 1980s the Packer group diversified its interest into mining and real estate with mixed success. In 1980 Consolidated Press moved into coal investment with a $24 million acquisition of the publicly listed mining company, Forestwood Australia Ltd. Subsequent diversification moves within Forestwood were the acquisition of a 51 per cent interest in Lefroy Salt Pty Ltd, and the takeover from Aberfoyle Tin Ltd of tin and wolfram operating mines, but with market crashes, mines were closed in 1982. In 1981 the group adopted a policy of property-development investment: investments have included joint ventures like the Gemini twin apartment building at Caloundra on the Sunshine Coast and the 345 hectare Burleigh Forest estate near Surfers Paradise, with Alan Bond's Walton Bond, and ski resorts in the Snowy mountains. Consolidated Press became one of the nation's biggest cattle breeders with the acquisition of three major Northern Territory properties – Victoria River Downs, Newcastle Waters and the Humbert River Station. Packer has also moved into oil, with holdings in the promising South Pepper – North Pepper oil fields in WA. Apart from mining and property interests, Consolidated Press has bought 75 per cent of mineral-water company Taurina Spa, a large stake in a major travel agency, World Travel Headquarters, a share in Lion Insurance, and a one-third share in Lotto Management Services which operates lotto gambling in NSW.

The remarkable growth of the group under Kerry Packer – 30 per cent per annum compound growth for eight years until 1982 – led him in 1983 to turn his attention back on the long-term investment prospects of the parent company itself. He moved to regain total outright ownership of the Consolidated Press empire. The Packer family company, Cairnton Holdings Pty Ltd, previously held only 47 per cent of the capital of Consolidated Press Holdings Ltd, but with a $114 million offer Cairnton Holdings gained full control of all of the ordinary capital of Consolidated Press. On the grounds of the massive leap in paper profits made by the conversion of Consolidated Press to full private ownership, *Business Review Weekly* escalated Kerry Packer's estimated private fortune to a conservative $200 million in 1984.

The Packer organisation has played a key role in the push for major technological changes within Australia's media systems. It was Packer's staff, particulary Les Free, executive vice-president of research and engineering, who provided the catalyst for investigation of the introduction of a domestic satellite in the late 1970s. The Packer organisation has also contributed more than any other media conglomerate towards the push for the greater privatisation of telecommunications in Australia. Changes in the commercial television industry have never been mooted at the expense of existing Packer commercial television holdings, nor have Packer staff supported innovations which would allow new players into the television system. At the Australian Broadcasting Tribunal hearings into the possible introduction of cable and pay television in 1982, Len Mauger, managing director of the Nine Network, presented a plausible case in support of the retention of the existing system. He attacked the case for the introduction of radiated subscription-television services on the grounds that 'there is no public demand for RSTV', which would 'bring nothing to the community other than an invoice' and that its fragmentation of the television service 'would have a detrimental effect on free television.'[22] While this position is obviously based on vested interest, the Packer organisation, of all major media groups, has generally presented the best researched and most professional submissions to government inquiries concerning new communications technologies.

Packer investments in technological diversification have been cautious and some plans to enter new markets have been scaled down or dropped. In videotex, for instance, Packer had been quick off the mark several years ago to secure the Australian agency for Prestel, the international viewdata service. He found, however, that it was virtually impossible for a private-enterprise organisation to set up a comprehensive network without Telecom. As he became wary of the market prospects of videotex, Packer expressed the view that videotex was an 'engineer's toy' with an uncertain future, and dumped it. Later, in 1981, Packer entered the home-video market with subsidiary company Publishing and Broadcasting Video, a broadly based video company with program acquisition, international distribution, manufacturing and home-entertainment distribution. However, in a strategy which may be encapsulated as spread the risk but not the rewards, Packer soon sold a 51 per cent public interest in a number of video divisions and subsidiaries for about $12 million as a huge sharemarket float in the local entertainment industry.

The Fairfax group is a well established large family business which began more than 150 years ago with the *Sydney Morning Herald* and is now a $500 million diversified media corporation. Members of the Fairfax family – James O. Fairfax, Sir Warwick Fairfax, Sir Vincent Fairfax and John B. Fairfax – control more than 50 per cent of the group John Fairfax Limited, which has virtually no interests outside of its substantial media investments. With an abnormally low corporate profile, the Fairfax organisation has maintained a fierce sense of independence in both its journalism and its corporate practices. Fairfax, for instance, unlike Murdoch and Packer, rejected an overture to participate with the NSW government in lotto operations, and refused for many years (until after the Norris inquiry) to join the Australian Press Council. To their eternal credit, the Fairfax organisation commissioned for the 150th anniversary of the

Sydney Morning Herald in 1981 an authoritative company history, *Company of Heralds,* and gave author Gavin Souter access to all company records and full editorial responsibility for the book. Australian media organisations prefer the veil of company secrecy and few insider accounts have been written about, let alone commissioned by, group media conglomerates.

The Fairfax group boasts an array of quality Australian newspapers as well as many of the nation's finest journalistic achievements. Its newspaper stable – the *Sydney Morning Herald, Age, Australian Financial Review, Canberra Times,* the weekly *National Times,* as well as Sydney's *Sun* and *Sun-Herald* – represents Australia's most analytical and influential press. The 83 per cent Fairfax interest (March 1985) in David Syme & Co Ltd, publishers of the *Age,* has added a specialist range of publications such as *Business Review Weekly, Personal Investment, Australian Pensioner, Video Age* and *Age Monthly Review.* Magazine Promotions Australia Pty Ltd has a diversity of magazines, with best circulation growth in recent years by *Woman's Day, People, Cosmopolitan* and *Dolly*. Fairfax also owns major regional daily newspapers, *Newcastle Herald* and *Illawarra Mercury* as well as a large collection of local community newspapers, primarily in Victoria. The group also owns 50 per cent of Australian Newsprint Mills (the other 50 per cent is HWT-owned), which produces more than three-quarters of Australia's newsprint. In radio, Fairfax's Macquarie network – 3AW (Melbourne), 2GB (Sydney), 2CA (Canberra), 2WL (Wollongong), 4BH (Brisbane) and 5DN (Adelaide) – has found a successful mix of personality-oriented serious public-affairs commentary and comprehensive sporting coverage. In television, its membership in the Seven Network – Fairfax controls ATN-7 Sydney and the Herald HSV-7 Melbourne – has not generated the strongly integrated Sydney-Melbourne axis of the Packer Nine and Murdoch Ten television networks. Late in 1983, Fairfax made a successful takeover bid for Brisbane TV Ltd, thereby gaining control of top-rating Brisbane television channel, BTQ-7.

Fairfax has shown in the past that it will go to extraordinary lengths to preserve the key balance of market forces within Australia's commercial media. Rupert Murdoch's aborted takeover of the HWT group in 1979 essentially failed because Fairfax was prepared to buy Herald shares well above any realistic market price to protect the Herald group – their competitor in many fields – from a News-group takeover. Had Murdoch won in 1979, a combined Murdoch-Herald conglomerate would have become Australia's biggest media giant, and one of Australia's top 20 companies.

Not only did it cost Fairfax an estimated $52 million to acquire a 14 per cent interest in the HWT group, a burden in interest repayments, but it was also obliged to sell its shareholdings in commercial television channels CTC-7 (Canberra) and QTQ-9 (Brisbane) to prevent any breach of the *Broadcasting and Television Act.* However, in the aftermath of the Murdoch raid, Fairfax sold its Herald shares to Queensland Press (39 per cent owned by HWT), thereby enabling it to bid for another television channel, this time BTQ-7 in Brisbane. Later in the mop-up, Fairfax moved to acquire greater interest in David Syme & Co Ltd when Ranald Macdonald, in an amazing volte-face, offered his and the remaining Syme family shares to the Fairfax group. Later, Fairfax offered to purchase the remaining 14 per cent shareholding of David Syme & Co Ltd held by the HWT. This was an extraordinary era of horse trading for shares among Australian media companies, and a real world game of monopoly for what are public assets – television licences.

The Fairfax organisation is the best placed of the big four to diversify its operation into the marketing of information services. Fairfax, publishers of papers of substance, have an established body of valuable information and can readily move into direct marketing of specialised information packages for particular users. Their acquisition of some overseas interests has shown an awareness of the potential of information-based organisations – *Asian Press and Media Directory, Computer Asia, Computer Asia Software Directory* and *Micro Asia.* With the takeover of David Syme, Fairfax acquired some valuable Australian subsidiaries which have made healthy profits from new technologies, notably AAV-Australia Pty Ltd, a Melbourne electronic-technology services and research company. The Australian Financial Review Information Service has been converted from a clippings service into *InfoLine,* a more broadly based data service for subscribers seeking company data and original research. In September 1984 Fairfax proposed a $96 million capital-raising venture, its first in twenty years. The group remained tight-lipped on the purpose of this substantially strengthened capital base, though the underwriter of the cash issue reported that 'the funds were to expand Fairfax's traditional business using new technologies relating to news and information services', particularly those involving the use of electronic and satellite communication techniques.[23] However, due to problems with the Sydney Stock Exchange, the board was forced to withdraw the share issue.

The Fairfax-controlled Channels Seven in Sydney and Brisbane have moved into television-delivered information services before all other competitors. In Feburary 1980 Fairfax launched Australia's first teletext system, Seventext, though Sydney's ATN-7. Teletext is a method of delivering information stored in a computer to domestic television sets, suitably adapted with a decoder to enable the set to display teletext information, and a keyboard, similar to an electronic pocket calculator, for the viewer to call up the pages of information. The information includes the latest news bulletins, sport results, financial news, rural data, TAB information, and travel, weather and community information services. This can be quickly updated and there is no direct payment to the channel by the viewer. Teletext has not been financially successful for many reasons, including the delays in getting equipment standards approved, the cost of decoders to consumers (about $500), the problems of developing user awareness, and the difficulty of attracting advertising revenue to such an unconventional television service. Teletext, known in the intensely profit-oriented television industry as Ted's Folly, after ATN-7's general manager Ted Thomas, has also been developed by Brisbane's BTQ-7 as Seventel. The staff of BTQ-7 have high expectations and their plans include the development of an ambitious hybrid teletex–videotex service that would broadcast special offers to viewers. With the recent acquisition of a Sydney-Brisbane television axis, Fairfax may upgrade their investment in this television-delivered information service.

Teletext is a technology which can be harnessed for considerable social benefit. For example, a closed captioning service enables deaf people to dial up television programs which have been sub-titled. The problem, however, is that sub-titling is an expensive labour-intensive process, costing about $7000 a week for a half-hour program. Commercial television channels have offered few sub-titled programs for the deaf; in Sydney during

1984 each commercial network offered only one sub-titled program (Nine's *60 Minutes,* Seven's *Sons and Daughters* and Ten's *Carson's Law)* while the ABC's Channel Two offered seven programs. The Commonwealth Government has funded the Australian Captions Centre, a national non-profit organisation which provides a teletex-based sub-titling service to the television industry for the benefit of Australia's estimated one million hearing-impaired citizens. However, a full service is dependent on the co-operation and support of the Australian commercial television industry which has been generally resistant to such moves. There is a cruel irony in this – the teletext system was originally developed for just this purpose, yet became sidetracked along the way into more commercial applications. To work in the context for which it was designed, teletext must have a socially aware institutional framework.

6 Public Broadcasting: Dilemmas of Direction

In the early history of broadcasting, the notion arose of a public-service broadcasting institution, a socially responsible organisation which made its broadcasting services available to the whole community. Following the establishment of national public-service broadcasting bodies in many democracies in the 1920s, those institutions developed a sense of self-pride and importance to their community and a need for their declared independence and protection from political interference. However, wherever broadcasting is organised as some form of public-service institution, subject to government regulations and legal statutes and heavily dependent for its revenue on the government of the day, the organisations are inevitably dependent upon political circumstance. The notion of a truly independent or autonomous public-service broadcasting institution is illusory.

There are, however, considerable variations in the way public-service broadcasting institutions are constituted around the world. Usually the distribution facilities, such as transmission equipment, are owned by the state. Methods of funding vary from those organisations which operate on revenue accrued from a licence fee, such as the BBC and NHK (Japan), where revisions of the licence fee charged must be approved by Parliament, to those almost solely dependent on government appropriation, such as the ABC. Others, such as the Canadian Broadcasting Corporation (CBC), are financed by a mix of annual government appropriation supplemented by advertising income. The form of management varies, from those organisations which may be classified virtually as orthodox public-service bodies, to those which are corporate bodies operating under the private law. Differences also exist in the manner of appointment of their respective heads; some public-service broadcasting organisations have a chief executive officer directly appointed by government, while in other organisations the head may be appointed by a Board of Governors who, in turn, are appointed by the government of the day.

Public-broadcasting organisations around the world face a series of contemporary crises. Their universal problems in the mid 1980s are those of inadequate resources, tensions associated with an uneasy sense of independence, and their crisis of identity. In recent years they have been increasingly analysed in narrow economic and industrial terms, and scrutinised in the context of a nation's macroeconomic strategy for growth and recovery. Where monetarists are in power, and where attention is focused on attempts to reduce deficits and slash public-sector spending, with top priority given to the hope of a private-sector-led recovery, public-service broadcasting organisations are natural targets for reduction. In those countries where the mood in contemporary broadcasting policy is towards deregulation and privatisation, such as the USA and the UK, public broadcasters, inherently dependent on public-sector financing, find themselves in an uncompetitive and threatened position. The traditional paternalistic defence of public-service broadcasting, invoking notions of programming substance and cultural worth, are widely perceived as being 'out of tune' with the times.

In the UK, relations between the Thatcher administration and BBC management have long been tense; the BBC has found itself with lack of funds, bitterly attacked for its programs about Nazi infiltration of the Tory Party, and absolutely unforgiven by the government for its 'unpatriotic' coverage of the Falklands war. With the recent approval of cable-television licences in the UK, and with direct-broadcast satellite systems probably on the way, the BBC finds itself caught up in the communications revolution. Not unexpectedly, BBC management has been somewhat preoccupied in recent years in coming to terms with what ought to be the role of the BBC in the face of unprecedented competition for audiences. The Adam Smith Institute Omega Report on *Communications Policy* typifies the mood for change in some quarters; this report attacks the 'outdated' broadcasting duopoly of the BBC and IBA in Britain, suggests the need for 'a most flexible and commercially aware licensing body', like the FCC in the USA, and calls for the dismantling of the BBC in its existing form.[1] In summarising the report in *The Listener,* Philip Schlesinger argued that 'given the accelerating drive towards privatisation in all sectors of the economy, it seems only a matter of time before calls for the wholesale dismemberment of the BBC move from the fringes to the centre of the political debate.'[2]

Broadcasting in the USA is dominated by the entrenched commercial networks – there is simply no real equivalent of the BBC or ABC. The Corporation for Public Broadcasting, established as late as 1967, is financed by meagre congressional allocations for its administration, as well as by corporate sponsorship for particular programs and by grants from Federal agencies and from private foundations. American public broadcasting is highly decentralised, with the Public Broadcasting Service (PBS) responsible for television, and National Public Radio (NPR) co-ordinating an impoverished network of local radio stations. Despite all of its problems with funding, and an increased dependence on corporate sponsorship during a period of economic recession, the PBS has managed to somehow keep a range of substantial programs to air, including America's leading nightly public-affairs program, The McNeil-Lehrer Report. Public radio only developed the semblance of a national programming network less than a decade ago, and its source of funds has never been secure. Under the Reagan administration, Congress was asked to completely stop financing NPR, but opted instead to drastically slash its funds. As a result, NPR found itself in dire trouble, resulting in staff dismissals and axed programs, at a time when its national audiences had trebled within five years, and when its flagship 90 minute current-affairs program, 'All Things Considered', was hailed as the most innovative public-affairs radio program in the history of American broadcasting.[3] The congressional squeeze in the USA, together with generally reduced support from the private sector and philanthropic organisations, threatens the survival of American public broadcasting.

Public Broadcasting: The Australian Way

Australia has a unique tripartite model of public broadcasting. The ABC is the nation's oldest and largest broadcasting institution: its 1984-85 Federal government budget allocation was $340 million; it employs 6800 full-time staff and 600 casuals and boasts the greatest number of broadcasting outlets of any Australian organisation.[4] A second public-service broadcasting institution is the Special Broadcasting Service (SBS), established in 1978, which is responsible for multi-cultural television services on Channel 0-28 in major capital cities and some ethnic radio services. The SBS budget allocation in 1984-85 was $39 million and it employs about 100 permanent officers and over 200 temporary staff.[5] The third member of public broadcasting is a collection of public or community radio stations – some 54 radio stations of diverse kinds, with a combined revenue base of a paltry $4.5 million annually, with 250 paid staff supported by 10,000 unpaid volunteers.[6] These community-based radio stations form the most impoverished sector of Australian broadcasting and their national fund-raising body, the Public Broadcasting Foundation, received a meagre $500,000 in the 1984-85 budget (though promised $15 million) of which $150,000 was allocated for grants to encourage Aboriginal programming on public-broadcasting radio stations with the balance of $350,000 for grants to stations.

The ABC has never been able to escape its history. The Australian Broadcasting Commission came into being under the Australian Broadcasting Commission Act of 17 May 1932. On Friday 1 July 1932 the then Prime Minister, Joseph Lyons, said in the ABC's first broadcast that the commissioners' 'task would be to provide information and entertainment, culture and gaiety, and to serve all sections and to satisfy the diversified tastes of the public'.[7] The then BBC Chairman, Charles Lloyd Jones, observed that the ABC was a fine set of initials for a body setting out to enlighten the nation, connoting the beginning of knowledge and 'forming an appropriate co-relative to the abbreviation BBC'.[8] The ABC's historical location is firmly within the tradition of its British counterpart and the BBC's declared pride in a commitment to truth and the pursuit of excellence in broadcasting was inherited by the Australian imperial artifact. There has long been a sense within the ABC of its own importance as a national institution and as a kind of central moral authority in Australian broadcasting.

Today, the ABC is a remarkable hybrid public institution. In radio, the ABC broadcasts on three major AM networks throughout 95 stations, as well as a 24 hour classical-music FM radio network; a Sydney-based rock station, 2JJJ; an overseas short-wave nine-language service, Radio Australia; and a short-wave radio service for outback Australia. The national television service, through 270 transmitter and translator stations, provides the most comprehensive geographical coverage of any television organisation in Australia, which will become even more extended in future years with inland community-broadcasting services via domestic satellites. In addition to these broadcasting functions, the ABC boasts six symphony orchestras, a national musical training organisation, the ABC Sinfonia, and a concert-management division which handles at least 750 orchestral and recital activities a year. The ABC also has a merchandising division, ABC Enterprises, which markets records, cassettes and books associated with its broadcasting, and the organisation is also involved with collaborative magazine publishing ventures such as *Look and Listen.* The facilities, diversity and potential of this communications infrastructure has been given little attention in discussions about the future role of the ABC in the Australian community.

Ethnic, or non-English-language broadcasting, developed separately from the ABC and commercial broadcasting organisations. Nineteen per cent of Australia's population is of non-English-speaking origin, though the figures are much higher for Sydney and Melbourne: 24 and 31 per cent respectively.[9] Responses to the broadcasting needs of ethnic communities were slow – commercial broadcasters offered a few token programs, usually on Sunday mornings, and the ABC has long remained a predominantly white-anglo-saxon-mono-lingual broadcasting institution. Consistent with government policy of migrant assimilation in the 1950s and 1960s, the Australian Broadcasting Control Board regulations limited foreign-language broadcasting to 2.5 per cent of stations' total transmission time. The formation of the SBS in the mid 1970s grew out of the needs and demands of ethnic communities, their increased politicisation, and the electoral pragmatism of Australia's major political parties.

It was the Whitlam Labor Government who set ethnic broadcasting in motion in July 1974 when they approved experimental licences for the establishment of subscriber-type services to provide specialist programs to the Greek and Italian communities in Melbourne and Sydney. In 1975 two government-funded ethnic radio stations were established in Sydney and Melbourne, 2EA and 3EA respectively. In September 1976 the Fraser Government invited the ABC to establish a permanent ethnic broadcasting service, but withdrew its offer in June 1977 and created an independent statutory authority, the SBS, which came into being from January 1978. The Galbally Report acknowledged the vital contribution that the extensive post-war immigration program had made to one of the most harmonious and cosmopolitan societies in the world, and it recommended the establishment of a multi-cultural television service in Australia.[10] This became a campaign promise by Malcolm Fraser in the 1977 Federal election campaign. After some delay, because Minister Staley could not get his plan through the Parliament for a more embracing statutory corporation, to be called the Independent and Multi-Cultural Broadcasting Corporation, the government fell back on SBS legislation, and Australia's multi-cultural television channel began its operation in Sydney and Melbourne on United Nations Day, 24 October 1980.

SBS has no equivalent in the world in its accommodation of a wide range of programs of many different cultures within limited broadcasting facilities. This has provided a valuable diversification of television in offering programs which would otherwise never have been seen. Genuine attempts have been made to accommodate a wide range of multi-lingual programming on television and the policy of using English sub-titles has increased their accessibility. However, SBS as a broadcasting institution has lived an uneasy existence.

Its prime problem has been that it has suffered from the lack of clearly enunciated government policies on multi-culturalism in Australia and its broadcasting responsibilities within that context. Commercial broadcasters, having ignored ethnic communities for so long, have attacked the organisation on grounds of its cost to the taxpayer and its duplication of 'proper' functions of the ABC, and have warned that ethnic television could result in 'electronic ghettos'. The PBAA has been critical that valuable television spectrum space is used for only about six hours a day, charging that the SBS, with squatter's rights to prime time, is 'a planning lapse of considerable magnitude'. The Connor inquiry into the SBS, established by the Hawke government late in 1983, demonstrated the considerable dissatisfaction within ethnic communities over the alienation of the SBS broadcasting services which are supposed to serve them. These general charges are that the policy, programming and productions are dominated by anglo-saxons, that SBS management imposes restrictions on community participation and that the organisation ought to be much more responsive to its communities' demands.

Public (or community) radio arose as a public-broadcasting service rather than as public-service broadcasting. In the first forty years of the Australian broadcasting system only two groups were allocated licences: commercial organisations and the ABC. Some 2000 requests for new broadcasting licences had been received and refused by 1970, and the convergence of a grass-roots radio movement and the mood of change which heralded the election of a Whitlam Labor Government in 1972 created this new sector of Australian radio broadcasting. The changes came, under Labor, in an ad hoc way, as part of a diffuse political notion of surrounding the commercial oligopoly with new broadcasters sympathetic to the Labor party, but the Labor Government public-radio proponents lacked a clearly articulated social strategy and failed to provide a secure legislative and funding base. The sources of requests for new radio licences included FM music enthusiasts, educational institutions who wanted their own avenues for programming, ethnic communities disenfranchised by the established anglo-saxon broadcasting monopolies and diverse local groups interested in direct community access for participatory and alternative programming. Radio 5UV-AM, at the University of Adelaide, was Australia's first public radio station in 1972, followed by 3MBS-FM and 2MBS-FM in 1974, stations of the Music Broadcasting Society of Victoria and NSW respectively. In Melbourne 3CR went on air in 1975 as the first community-radio station supported by a group of diverse community interests.

Contrary to fears about their future under the Fraser administration, Minister Staley gave the public-broadcasting sector a secure legislative foundation. Amendments were made to the Broadcasting and Television Act in 1976 and three types of public-broadcasting stations were defined: category E licences for educational bodies, category S licences for groups intending to 'serve a particular interest or group of interests', and category C licences to 'community groups intending to provide programs serving a particular community'. Public-radio enthusiasts were so swept up in their haste to get stations on air that the question of how they were to be adequately financed was virtually ignored. A new sector of Australian broadcasting was born, but without finance from conventional commercial advertising and with no assurances of any government funding. Lack of adequate guaranteed income has remained an endemic problem to public broadcasting. Finances are raised by listener subscriptions, on-air fund-raising appeals, short paid announcements, sponsorships from organisations such as unions and tertiary institutions and meagre grants from State and Commonwealth governments. The energies of the few paid staff tend to be sapped by fund-raising drives and the overall effectiveness of public radio has been severely constrained by inadequate resources.

Despite these problems, Australia now has a remarkable array of public-radio stations. It is not only the different kinds of station – rock, classical music, ethnic language, radio for the print-handicapped, Aboriginal stations – but the localised and alternative program contents explored: trade unions, senior citizens, media watch, prisoner and welfare and community activist programs. The openness which characterises most of these stations, where ordinary people are given access, has meant that public radio has emerged as the most responsive and participatory sector of Australian broadcasting. In a small step towards the democratisation of the airwaves, the notion of slick golden voices of commercial radio or the authoritative tone of ABC announcers has been overturned in favour of ordinary mortals speaking on radio with their natural voices. However, the perceived unprofessionalism of voice, or rawness of production, has alienated many listeners who have not been able to accept presentation on radio other than in conventional modes.

The ABC: Entertaining Changes

In the long history of the ABC, the last decade has been its most traumatic. The Fraser government came to power in 1975 committed to the rhetoric of small government and a private-sector-led recovery, with a senior cabinet deeply hostile towards the ABC and a Prime Minister who was to become more involved in the affairs of the ABC than any of his predecessors. Basic changes to the structure of the Australian broadcasting system were mooted, and in particular that the ABC would be made much more accountable for its actions with the creation of one supreme broadcasting authority to which both the commercial and the national public sector would be responsible. A report by Fred Green, Secretary of the Department of Postal and Telecommunications, recommended in September 1976 the disbandment of the Australian Broadcasting Control Board and its replacement by an Australian Broadcasting Tribunal with greater regulatory authority.[11] Fraser's appointment of Bruce Gyngell, a former managing director of commercial channels Seven and Nine (against the wishes of the responsible Minister), as the inaugural Chairman of the Australian Broadcasting Tribunal in January 1977, symbolised the Fraser government's intentions in broadcasting policy. ABC management, however, were able to out-manoeuvre their minister and resist attempts to put the national broadcasting organisation under the wing of the newly created authority.[12]

The highlights of the early part of the Fraser Government's war of attrition on the ABC are well documented elsewhere.[13] In their first few months of office the Fraser Government proposed to reduce the sum appropriated to the ABC in the Budget session of Parliament by $7.1 million to $125 million.[14] Acting Chairman, Dr Earle Hackett, fought the proposed cuts, warning that more than 1000 staff would have to be retrenched, a situation he regarded as tolerable only if the nation was judged to be in a crisis comparable to war. Dr Hackett's replacement was Sir Henry Bland, former Secretary to the Department of Defence, sent in to 'clean up the place', and who was clearly hostile to the values of the national broadcasting organisation he led. Financial cuts to the ABC during the Bland era led to the cancellation of proposed drama series, cost-cutting in highly regarded public-affairs programs such as *Four Corners* and *This Day Tonight,* abandonment of the recruitment of specialist trainees and a serious loss of staff morale. Bland resigned in December 1978, charging Fraser with broken promises about prior assurances given concerning the number of ABC commissioners in the revised Broadcasting and Television Act, to be replaced by John Norgard, a former executive of BHP. In the 1978-79 budget the ABC's funding was reduced by 3 per cent in real terms and the commissioners resolved that Norgard should report to the Minister that the government was 'seriously prejudicing the commissioners' ability to carry out its basic statutory responsibility.'[15]

This era of contrived hard times for the ABC shows the manifestations of strict monetarism on a public institution. Under the Fraser Government the ABC was a broken, dispirited organisation with a considerable loss of self-esteem, producing an array of generally bland and timid programs. Moreover, this was at a time of abnormal healthy buoyancy for the ABC's counterparts in commercial television and radio. Later in his term of office, however, Fraser tactically retreated from crude ABC bashing, partly in response to public outcry, and his cabinet came to explore other possible methods of change to the ABC.

Rather unexpectedly in May 1979, the Minister for Postal and Telecommunications, Tony Staley, announced a full-scale inquiry into the ABC. The Committee of Review was to examine the 'services, policies and performance of the Commission', including 'programming policies and procedures for the maintenance of program standards', funding arrangements and management systems, and also to recommend on 'resource allocation practices and planning for technological developments'.[16] Alex Dix, managing director of transnational Reckitt and Colman Australia Ltd, was appointed head of the inquiry, together with committee members, law-Professor Alex Castles, film-producer Patricia Lovell and businessman Brian Sweeney. Given Dix's managerial position, and his experience in analysing corporate conglomerate problems, the focus of investigation was inevitably to be on the corporate management strategies of the ABC. Dix engaged consultants to review the financial management, industrial relations and staff management, engineering and technical services, as well as studies of ABC audiences and ABC records.

The 18-month Dix Review received 2259 submissions from 3500 people, held 61 hearings in 44 places from Port Hedland to Arnhem Land, and met 831 witnesses who gave 5500 pages of testimony. Their commendable five-volume report, *The ABC in Review: National Broadcasting in the 1980s*, was presented to the Minister of Communications in May 1981 complete with 273 sweeping recommendations. The essential thrust of the report was that top priority must be given to changing the management structure of the ABC. Dix recommended that the existing Commission be replaced by two bodies: a Board of five to seven directors, chosen by the government to make policy; and a National Broadcasting Consultative Council, of perhaps 20 people, chosen by Federal and State governments, to represent community views and provide advice to the Board. There was some evidence that the final recommendations concerning the desirable level of management shake-up were not as radical as Dix himself would have preferred. Apparently, Dix wanted to propose a spill of all ABC staff positions, about 6000 jobs, and was keen to recommend the same for the Special Broadcasting Service. It was reported that the Minister, then Ian Sinclair, did not demur at the prospect, but Fraser said no.[17] The problem with such a pragmatic approach by the head of a public inquiry is that government reports, unlike politicians, remain for posterity.

Dix was highly critical of many aspects of the performance of the ABC though he acknowledged the efforts of many talented and dedicated people who worked for it, and the general high esteem with which the community regarded the organisation. Dix argued that the ABC had 'faltered in the last decade' and 'stood on its dignity and independence when pressing priorities cried out for attention'. Instead of being the innovative broadcasting pace-setter expected of a national public-broadcasting institution, Dix charged that the ABC 'has not only slipped from the forefront of change but threatens to be eclipsed by it.' His summation – that the ABC had become 'slowmoving, overgrown, complacent and uncertain of the direction in which it is heading' – became slogans of onslaught for Australia's biggest public-broadcasting institution.

Much of Dix's report showed how he was swept up in the climate of monetarism. Essentially Dix argued that the problems he perceived in the ABC could be remedied by the introduction of a new structure of management and by increased efficiency. Rather than offering any criticism of government cut-backs to the ABC, he was in fact more critical of the ABC's response to cuts. He did not argue that the ABC be left to make its own decisions about staffing levels within the approved budget, but that government-imposed staff ceilings should actually remain until the ABC had demonstrated more effective control of its financial resources. There was much evidence, Dix charged, to suggest that 'the ABC has not managed its finances as well as it might have', and reference was made to 'worldwide trends to keep a rein on broadcasters'. ABC management was urged to generate more funds from internal activities and to also look outside their organisation for additional sources of funding. Better financial management systems could be introduced to the ABC, according to Dix, for a cost of about $6 million. What pervaded the Dix

Committee of Review was a drive towards efficiency and coming to terms with what they perceived to be the economic realities of the 1980s. Their report displayed limited understanding of the philosophy of public broadcasting. Notions of change and innovative programming possibilities unrelated to costs, budgets and efficiency appeared not to be understood.

The Dix investigation must be seen in the context of a drive towards privatisation of Australian institutions; Dix made it clear that 'the objective of many of our recommendations is to structure the ABC more like a commercial enterprise'. Yet this impetus lost its momentum towards the end of the Fraser administration's term of office, the organisation was less under siege from Canberra, and government took an inordinate amount of time to come to terms with the Dix Report. Minister Neil Brown announced the government's response on 4 July 1982 (Dix had reported in May 1981) and complimented Dix for his presentation of 'one of the most painstaking and thorough inquiries in recent Australian history.'[18] Brown announced the government's intention to provide the ABC with a charter to determine the organisation's future direction and he foreshadowed the establishment of an Australian Broadcasting Corporation. Dix's proposed revamped management structure was essentially accepted – a board of directors presided over by a chairman and a managing director as ex officio member of the board with full voting rights. The need for the board to be supported by a formally constituted advisory council was also accepted in principle. There was enough positivism in Brown's 32-page press statement to indicate that, if the ABC was to be put down some day, at least it would not happen in that year, the ABC's fiftieth year of operation. In fact the Minister offered the remarkable comments in view of the immediate past history of relations with the ABC that the Fraser government considered that 'the ABC has much to offer the nation' and that it wanted a 'vigorous, outward looking confident national broadcaster.'

The Dix Report had received bipartisan support and soon after the election of the Hawke government in March 1983, the Communications Minister Mr Duffy declared that the government 'wanted to make haste with the reorganisation of the ABC.'[19] The Australian Broadcasting Corporation Act, with slight modifications to the Liberal Party plan, passed through the Parliament in May 1983. The ABC was to be headed by a managing director who would be appointed by and accountable to six to eight non-executive board directors, including a chairperson, deputy and staff-elected deputy. The ABC was released from control of the Public Service Board and was expected to act like a corporation, though the restrictive clauses on investment and borrowing remained. The new Act did not really provide for fundamental changes to the ABC, but was merely a management systems overhaul. However, the publicity which surrounded the Labor government's clean-sweep board of directors generated a sense of change and seemed to herald a new era of opportunity for the national broadcaster. Labor chose mercantilist Ken Myer, the elder son of Sir Sidney Myer, founder of the retail empire, as the inaugural chairperson of the national public broadcasting corporation. It was a board of talented and imaginative people, but with limited experience in broadcasting and virtually none in broadcasting policy.[20] Within a few months of its formation the board chose British-born Geoffrey Whitehead as its managing director. His background was in political journalism and he was former director-general of Radio New Zealand. The new appointments, well publicised in the national press, brought with them a sense of a renewal for the ABC.

The press honeymoon for the new board was shortlived. After banner headlines in mid-1984 about chairman Myer's support of a management decision at the request of the government of New Guinea to ban a *Four Corners* interview with a rebel Irian Jaya leader and the board's decision to reverse that position, there emerged a series of leading feature articles and editorials as damning indictments of the new board. The *Age* (editorial 10/9/1984) claimed that 'the corporation found itself in trouble' at a time when it seemed the ABC 'was about to undergo a change of heart and purpose'. The Melbourne *Herald* offered two long feature articles where the disenchanted Alex Dix and deposed chairperson Professor Leonie Kramer said what they thought had gone wrong. An accompanying editorial alleged that the new board seemed 'to concentrate their energies on matters of monumental pettiness' (12/9/1984). The *Australian* charged that 'at no time in its history has there been so much public dissatisfaction with its performance, nor has there been so much uncertainty and gloom about its performance' (18/8/1984). The *Australian Financial Review* (3/9/1984), wearing its monetarist blinkers, attacked the new ABC on the grounds that 'large sums of money are being wasted on this moribund organisation.'

The early universal condemnation of the new board of the ABC by the popular press was an important component in fashioning public consciousness of the national broadcasting institution. It suits the interest of commercial media to continue the general impression that the ABC is down and out, yet again. The ABC, rather like the ALP in the 1950s and 1960s, tends to have its best rows in full public glare, a process facilitated by its own journalistic hawks who are less subject to self-censorship than their counterparts in commercial media. By contrast, the boards of commercial media organisations, with their faceless directors, are never hounded by commercial journalists in search of stories, and there can be virtually no public debate about policy formulation within Australian commercial media institutions. We also seem to generally expect higher standards of the ABC than we do from commercial media organisations. It is a glib response to argue that the ABC is funded by taxpayers – what must be remembered is that the community pays for the broadcasting system as a whole.

The new board of the ABC has been slow though to realise its promise and tardy at the implementation of change within the organisation. How could a completely new group of part-timers, previously unknown to each other, largely ignorant of matters concerning contemporary broadcasting issues, and who inherited a policy vacuum within the ABC, be instant catalysts for a revitalised institution? There seemed to be an assumption on the part of Australia's major political parties that if they changed the broadcasting act, renamed the organisation as a corporation and made a complete clean sweep of board personnel, a better ABC would somehow emerge. The ABC requires a board which understands its operations, which has continuity of membership, and which is given the opportunity to develop a long-term integrated policy strategy for public-service broadcasting in the future. The new board have sensed the need for change, but they have been unsure about the form and substance of those changes.

As a response to constant criticism in commercial media, the ABC initiated a major advertising campaign late in 1984 as an attempt to re-assure all that 'the ABC does have a very different direction'. In large advertisements (e.g. *Australian* 13/10/1984), captioned 'the ABC explodes a few myths', the organisational goals endorsed by the new board were highlighted. The ABC, we were told, will:

★ Strive to be a valued and vital part of Australian life.

★ Explore major contemporary social, economic and political issues as well as development in science, technology, education, philosophy and religion.

★ Encourage the appreciation and performance of the arts.

★ Work with relevant regional, ethnic and indigenous groups to foster appreciation of Australia's unique and diverse community.

★ Provide a focus for understanding Australia's place in a changing world and project Australia to the world.

★ Maintain its integrity and independence.

Few could disagree with such admirable statements, but the subtext for change and new directions was understandably missing.

What must happen in the ABC is a process of debureaucratisation and the devolution of power within the institution. A pyramid management hierarchy, with a general manager at its head and with only limited authority for the regionals and program-makers, is no longer appropriate for the organisation or the community. Power needs to be much more diffused within the ABC, with the development of a flat or horizontal management structure where the locus of power is eventually held by centres of programming. The new board know that they will eventually be judged on how successfully they shift balances of power within the institution. The change of dividing the organisation into nine divisions is a step in this direction – Television, Radio, Radio Australia, Information Services, Engineering and Property, Human Resources, Finance and EDP, Sales and Marketing and Concert Music. The ABC needs much greater mobility in its top management to get new blood and fresh ideas into the system, and to break the notorious old-boy network of senior management who so successfully lobbied the commission to get what they wanted. The announcements by the managing-director in February 1984 that senior managerial appointments would become fixed terms of office rather than be virtually life-tenured, and that more than one hundred ABC management positions would be open to outside competition within a year, indicated promising intentions of change.[21] A leaner ABC management is desirable, where the resources are essentially placed in the hands of program-makers who control their own budgets and are responsible for them, and who are more centrally involved in the major policy-making decisions of the organisation. Moreover, management and program development must be broadened to include wide representation from user-interest groups.

A contemporary problem for public-service broadcasters has been that their programming services have become more widely criticised and challenged, and they have usually failed to defend themselves with a clearly articulated programming philosophy. One level of attack is that ABC programs supposedly fail to capture adequately sized audiences. Divisions of audiences into 'mass' and 'minority' markets are gross generalisations, and conventional ratings surveys have been used to perpetuate the myth of the ABC playing to a small, educated, intellectual elite group. What the Dix Committee showed, based on their commissioned research of the ANOP National Population Study, was that 'the real minority is the seven per cent of people who neither watch ABC television nor listen to ABC radio within any given week.'[22] Conventional ratings surveys, which merely show the market share of the different programs at a given time, are purely commercial phenomena used to assist advertisers in their commercial negotiations and for commercial broadcasters to set their advertising rates. Conventional ratings do not take into account the form or substance of programs, audience responses, or their sense of appreciation of particular programs. Specified vested interests, such as commercial broadcasters and advertising lobbyists, maintain a dialogue which seeks to categorise the ABC as a wasteful and ineffectual component of the broadcasting system on the grounds of its inability to capture large audience numbers across the board.

The ideals of public broadcasting need to be reasserted. A primary role of a public-broadcasting organisation is to present a wide range of programs which offer intelligent interpretations of the community they serve. The ABC's role is to make programs for the heartland of the broadcasting audience, to address itself to the issues which are of concern to the community at large, and in this context the notion of minority audiences is meaningless. In public broadcasting, the criteria involved in deciding whether a program should be made are not those of a crude game of guessing a proportion of total audience shares. Programs are made because of a notion of their inherent worth, though often difficult to articulate, rather than a crude application of straightforward commercial criteria. Public broadcasting also seeks to address itself to the public in all its diversity. Unlike commercial media, it attempts to satisfy a wide range of different public tastes, not only in spoken-word programs, but in music, drama and other cultural tastes. As Stuart Revill of the ABC has put it, the ABC 'doesn't divide the audience into the 18s to 24s; or the two income family; or the A-type socio-economic; or up-market or down-market. It aims to discover, evoke, foster and satisfy at some time all those interests that are mixed into each of us and which give us our precious individuality.'[23]

The ABC needs to develop greater editorial courage to defend itself if power in the organisation is really devolved. The function of a public broadcaster ought not be to mobilise support for the existing order, but to offer exploratory programs which challenge the assumptions of the audience, and which may sometimes antagonise or offend listeners and viewers. ABC management in the past has been too concerned about reactionary attacks on programs such as *Broadband, Lateline, Doubletake,* and *The Coming Out Show* – labelled as prejudiced, biased or overradical – and too little concerned with defending the role of the ABC as a challenging program-maker. The organisation has been too closed and too undemocratic for too long – much could be learned from the community access and public participation offered by the public radio stations in recent years. If the ABC evolves as a less paternalistic and more open and accessible institution to a range of new program-makers, then it will have to learn to cope with the political flak which comes with provocative and contentious programming.

7 Telecommunications Institutions: Pressures for Change.

While Australian institutions tend to be categorised either as a public utility or a private enterprise, there are considerable variations within public ownership: there is a tax-paying corporate entity, such as Qantas; a non-tax-paying commission, such as Telecom; or a business organisation which is a department of the public service, such as the Australian Government Publishing Service.

In Australian telecommunications, common-carrier status is shared between three public statutory corporations: Telecom; the Australian Overseas Telecommunications Commission (OTC); and the national satellite-system authority, Aussat Pty Ltd. In 1975, following the Vernon Inquiry, domestic posts and telecommunications services were divided between the Australian Telecommunications Commission (Telecom) and the Australian Postal Commission (Australia Post), with OTC retaining its control over international telecommunications services, notably international phone calls.[1] In keeping with overseas institutional arrangements at that time, postal and telecommunications services were functionally separated, to be run as statutory corporations, or commissions, rather than government departments.

In 1979, the Fraser Government decided that Australia was to have its own domestic satellite system and that it was to be controlled by an authority separate from Telecom. This was a fundamental institutional change in Australian telecommunications and it established dual roles for public enterprises in the domestic telecommunications-services market, inevitably an uneasy alliance. In November 1981, following a short period of interim management by OTC, the Commonwealth government established Aussat Pty Ltd, a company incorporated in the ACT with sole responsibility for the operation of Australia's domestic-satellite system. Under the Fraser administration, the intention was that Aussat Pty Ltd would eventually be listed on Australian stock exchanges, with up to 49 per cent of the issued share capital available for private ownership. Soon after its election in March 1983, the first Hawke government decided that the Commonwealth Government would remain as sole shareholder in Aussat Pty Ltd. Later in 1983 it offered a 25 per cent shareholding in Aussat to Telecom, which was subsequently taken up by the terrestrial carrier. A strong body of opinion within the Australian Labor Party Caucus narrowly failed to have Aussat fully amalgamated as part of Telecom. Hence, there is now bi-partisan agreement for the development in telecommunications of an alternative domestic common carrier to Telecom.

The Old Firms: Telecom and OTC

The principal institutions of the Australian telecommunications system, unlike the predominantly private character of Australia's major media organisations, remain public in structure and character in the mid 1980s. The role of statutory telecommunications corporations is essentially to facilitate transmission services of different kinds as national common carriers. In assessing these institutions, what must be remembered is that their prime functions are about *carriage* and not content. The processes of technological change threaten to place great stress on existing institutions with the reduction or possible elimination of what was originally perceived to be a natural monopoly.

Telecom is Australia's largest and most profitable public corporation – the phone company Australians love to hate. Telecom has long benefited from a legally guaranteed monopoly until the establishment of a national satellite system. The *Telecommunications Act* 1975 vests authority in the Australian Telecommunications Commission (Telecom) to plan, establish, provide, operate and maintain telecommunications services within Australia 'as will best meet the social, industrial and commercial needs of the Australian people.' No person may construct, maintain or operate telecommunications installations within Australia, unless authorised by Telecom or otherwise permitted by Commonwealth statutes: the *Telecommunications Act* 1975, the *Wireless Telegraphy*

Act 1905 or the *Broadcasting and Television Act* 1942. No person may attach a line or piece of equipment to the public switched telecommunications network unless authorised by Telecom.

The existence of this monopoly has facilitated the development of a successful public corporation: in 1982-83 Telecom's operating revenue amounted to $3636 million for a trading profit of $263 million, with a full-time workforce of 88,524 people.[2] The $12 billion terrestrial-telecommunications network, progressively extended over past decades and technically highly regarded by world standards, is Australia's most valuable public asset.

Telecom's profit level, usually in excess of $250 million annually, is a contentious issue. As a public enterprise, Telecom has a responsibility to provide services not merely in response to the maximisation of profit, but to provide services in response to need. Section 6 of the *Telecommunications Act* requires Telecom to 'perform its functions in such a manner as will best meet the social, industrial and commercial needs of the Australian people and make its telecommunications services available throughout Australia for all people who reasonably require those services.' It is somewhat difficult to come to terms with the profitability of a statutory corporation, especially one with no share capital and which is exempt from all taxes and makes no payment of a dividend, as is the case with Telecom. The organisation is not, of course, free to determine its own level of overall profitability, and it is subject to government controls on its level of borrowings and for the approval of its basic tariffs.

Telecom mounts a substantial capital expenditure program each year for the maintenance and extension of its network. During 1984 it committed $1670 million to its capital expenditure program.[3] In theory if it wound back this capital program, reduced its level of borrowings, and increased its charges, it could produce a gigantic overall annual profit. However, its capital-expenditure program, at least 50 per cent of which must be internally funded under Section 73 of the *Telecommunications Act* 1975, provides substantial economic benefit at a time when private-sector capital investment is abnormally cautious. What must be kept in mind in a discussion of Telecom's profitability is its public character, in that the Australian community are its shareholders.

Telecom's major services of telephone, telex and data are significant components of Australia's information economy. The steady growth of new telephone connections throughout Australia puts her amongst the highest connections per head of population in the world. Despite Australia's sprawling geography, Telecom's telephone connections in the past quarter century have expanded from 700,000 in 1950, to 4 million in 1977, to 5 million in 1980 to in excess of 6 million during 1984.[4] Almost every business, and about 80 per cent of homes, have access to a telephone service, with metropolitan areas approaching saturation. Telecom management, in keeping with most other PTTs, has accepted the economics of integrating digital switching and transmission techniques for developing the telephone network towards an integrated digital network (IDN). Customers in capital cities will have first access to this all-digital technology some time in 1986. This system relies on a new generation of computerised telephone exchanges, called AXE, purchased from the Swedish company, L. M. Ericsson, which functions exclusively on computer technology to switch telephone calls. Telecom management has a declared objective to introduce a fully automatic national telephone service by 1990. The policy of progressive extensions of computerised telephone exchanges has profound implications for Telecom's workforce and has contributed to major industrial disruptions in the late 1970s.

Since its formation in 1975, Telecom has been progressively transformed from a basic telephone company to a more broadly based telecommunications-network service provider. This change has brought with it some questions regarding the directions of its development. Telecom management has to decide whether to pursue the goal of providing every dwelling in Australia access to the telephone network, or whether to invest more in the highly profitable east-coast business services or whether to diversify into as many different new telecommunications services as quickly as possible. In recent years Telecom management has adopted a more mercantile ideology. The organisation has never lacked its share of critics, from household customers who complain of long service delays to television executives outraged by breakdowns in interstate television bearers. There is a widely held view in the business community that the bureaucratic giant has been too slow to adjust to new prospects in information technology and is generally too unresponsive to customer requirements. Some see these problems as an inevitable carry-over from what was an essentially engineering organisation in the Post Master General's department. There has been a gradual transition towards a much more competitive marketing institution in an era of information services.

Telecom has moved rapidly into new telecommunications-service areas. Its biggest new growth area is the data-transmission field where business demand has quickly outstripped Telecom capacity. Two new data-transmission services, Digital Data Service (DDS), a leased circuit service, and Austpac, a new digital packet switched data-transmission service, were launched in December 1982. DDS was intended to replace Telecom's Datel Service as the main leased-line service which offers a basic range of point-to-point and multipoint data services. Austpac has the capacity to interconnect different computers and terminals, using either the public switched telephone network or private leased lines. Telecom's oldest telecommunications service, the public telegram service, run by Australia Post as its agents, is continued as a public service despite losses every year. A public telex service, either through Telecom business offices or via equipment on the premises of a telex subscriber, has enjoyed steady growth in recent years. Other service innovations are Teletex, the revolutionary mail transfer system, and Viatel, the ambitious new national videotex service. A decade after the French PTT introduced its extensive national data service to home and business customers via the public network, Telecom is offering its contribution towards Australian telematic society.

The Overseas Telecommunications Commission (Australia) is the only public authority in broadcasting or telecommunications that has not been subject to a national government inquiry during the 1980s. OTC is a statutory authority established by the Overseas Telecommunications Act 1946, with responsibility for the establishment, maintenance, operation and development of public telecommunications services between Australia and other countries, Australia's external territories, and ships at sea. Australian governments have chosen to institutionalise the split between domestic and international telecommunications: the Vernon commission into telecommunications reviewed this in 1974 and on the insistence of chairman Vernon, OTC retained its separate identity. OTC's financial performance, particularly in most recent years, has been impressive. Moreover, tariffs have remained unaltered over many years which has represented a cost reduction in real terms to the user. For the year 1982-83, revenue was $300 million, for a profit in operations of $42.7 million, with a total staff employment of 2486 people.[5]

OTC is a body corporate, wholly owned by the Commonwealth. The Commission pays income tax and conducts its affairs in accordance with normal commercial practices. Over the years Treasury has regarded OTC as a favourite milch cow, with a resultant reduction in funds available to the Commission for investment. Chairman Somervaille complained that in the five-year period from 1977/78 to 1981/82 OTC had a total profit of $108 million after repayments to the Commonwealth Government of $97 million in income tax, $94 million in dividends and $18 million as repayment on capital.[6] As a consequence, OTC has run a significant working-capital deficit and assumed substantial long-term debt. Whilst OTC's annual turnover is only a fraction of Telecom's, its profit growth as a percentage of revenue is generally the best of Australia's nine major statutory corporations.

OTC services have shown steady expansion in recent years, with high growth in its international telephone-service area, which generally contributes about 70 per cent of total revenues. OTC provides an international telephone service to more than 280 countries, with about 130 of those countries now available through ISD, international subscriber dialling. OTC, of course, merely plays a carrier role in international communications, and must work in close liaison with international bodies: Intelsat, the International Telecommunications Satellite Organisation and Inmarsat, the International Maritime Satellite Organisation. As well, it coordinates services with Telecom, which is responsible for calls internally until they reach OTC's gateway terminals. OTC and Telecom have a five-year agreement on ISD financial arrangements: Telecom, in 1984-85, charges OTC a fixed figure of 33 cents per minute for ISD calls, but this charge will drop to 31 cents in 1985-86 and 29 cents in 1986-87.[7] Some 40 per cent of OTC's collected revenue generally goes towards payment of domestic reticulation costs.

There are considerable industrial-relations tensions within OTC operations, where the moves to reduce operator-connected telephone calls gathers strength each year. OTC management, striving for better returns and concerned about increasing labour costs, wish to encourage direct consumer dialling as much as possible. Late in 1983, OTC announced increases on a number of international telephone charges, including a $2 surcharge on those operator-connected calls which could be made by direct dialling. Understandably, the Australian Telephone and Phonogram Officers' Association expressed their displeasure at these moves, and the Australian public has, on occasions, enjoyed free international phone calls from some operators. In the long run, OTC's management approach to the increased use of international subscriber dialling is likely to lead to a substantial loss of telephonists' jobs.

Like Telecom, OTC has in recent years diversified its range of international telecommunications services, though it lacks the comparable monopoly protections of Telecom, and it faces stiffer competition from international private entrepreneurs. Examples are OTC's two electronic messaging services, MIDAS (Multimode International Data Acquisition Service) and MINERVA. Midas, launched in 1979, provides customers with access to computer data banks overseas, using a terminal incorporating a keyboard, with a display screen or printout facility. Its 2000 or so registered users include corporate-government, educational and research bodies who are able to call up highly specialised information instantly from data banks in several countries. Minerva is an extension of the Midas data service, designed as an upmarket service to enable executives to handle correspondence by using a keyboard connected to the telephone network. This is an electronic messaging service where the executive can view any mail lodged in the 'electronic in tray'. OTC also offers two forms of fixed private services, International Leased Circuits, where almost 500 customers lease private lines to link their Australian office 24 hours a day with overseas offices, and Interplex, where about 50 customers lease a portion of OTC's computerised switching system to create a private international switched network. Meanwhile, more humble services, such as the international telex service, show a steady growth but there is a growing trend away from the use of international telegrams in favour of alternative forms of communication.

OTC also offers a high-speed document-facsimile service, called Overseasfax, which transmits printed, typed or handwritten documents to an overseas bureaux address. Intertel is a service designed for people in small businesses to send written messages to overseas correspondents by telex, facsimile or telegram. OTC's biggest capital-investment involvement is with the ANZCAN (Australia-New Zealand-Canada) cable system. ANZCAN, completed late in 1984, spans the Pacific, extending from Sydney to Vancouver, via Norfolk Island, Fiji and Hawaii. This $300 million, 14-nation collaborative venture provides submarine cable circuits to link Australia, New Zealand, Fiji, Canada and the USA. In an era when satellites are seen as the glamorous newcomer to the international telecommunications stage, this ambitious submarine cable venture received little publicity. This typifies OTC's efficient, low public profile to international communications.

Pressures for Privatisation: Alternative Telecommunications Networks

The buoyancy of telecommunications markets in developed countries in recent years has naturally attracted the close attention of astute private-sector entrepreneurs. Who would have contemplated when the original PMG telephone system in Australia began as a useful public utility that by the mid 1980s telecommunications organisations would be highly attractive to speculators looking for investment prospects? The pressures from the private sector to change public ownership and control of telecommunications networks is a contemporary economic phenomenon in most developed countries, and, in Australia, both statutory corporations, Telecom and OTC, are subjected to forces for change.

These arguments were most forcefully put by the big private-enterprise communications lobbyist Business Telecommunications Services, whose original membership included IBM, AMP, ACI, Ampol, AWA, BHP, CRA, Myer, PBL and TNT. Business Telecommunications Services was established in September 1980 with an announced brief to 'research and define' the long-term communications needs of business. But it has emerged as both dedicated to the diminution of Telecom's common-carrier role and as a prospective operator for new services if the market is deregulated. Its general manager, Mr Peter Holmes a'Court, formerly of IBM, argued that the principal concern which led to the formation of BTS was the unfortunate dependence on Telecom. Speaking at ANZAAS in May 1981 he said that, above all, the genesis of BTS was 'fuelled by the painful and costly experience of two national telephone strikes within a period of 12 months – an acute sensitivity to the fact that business (indeed the entire community) is almost totally dependent upon Telecom, as the sole source of supply for this most vital of commodities.'[8]

Australia's telecommunications network has been crippled several times in recent years by the strike actions of Telecom's key technical union, the ATEA. Disputes which led to major shutdowns of the communications network occurred in August 1978, August 1979 and June 1981, for periods of two to three weeks. In the 1981 strike, operator-connected interstate telephone calls had progressively longer delays until eventually most STD services became inoperable, the international subscriber dialling was non-operational, many business organisations lost computer data-link contact between interstate branch offices and private-line networks leased from Telecom were down. Telecom workers clearly demonstrated how effectively they could disrupt the nation's communication systems, at considerable cost and inconvenience to the Australian public. The ATEA was labelled an irresponsible strike-prone union and the 'communications chaos' arising from dependence on a single integrated terrestrial system emerged as an issue for the future direction of Australian telecommunications policy.

Any fair-minded assessment of these arguments by the BTS consortium must take into account that industrial conflicts were not merely confined to the telecommunications industry and that arguably Telecom disputes of the late 1970s featured many of the worst elements of the Australian industrial-relations system: inept management, a dogmatic government, an inflexible arbitration and conciliation system, as well as bloody-minded unions. Bill Mansfield, secretary of the ATEA, responded to the accusation that Telecom was in the hands of a 'strike-prone' union, by arguing that there had been only three major industrial disruptions to Telecom due to employee industrial action in its 80 years of service to the community. Moreover the causes, rarely analysed in media presentations of the disputes, needed to be understood. Mansfield explained: 'It should be noted that the industrial action which has taken place was caused, firstly, over the widespread and rapid introduction of new technology and secondly, because wage levels of skilled trades and technical employees were considerably below community standards for similar work.'[9]

Central to industrial disputation within Telecom was the introduction of new technology. Management had decided to introduce two new forms of computer equipment into telephone exchanges, known as ARE 11 and AXE, which Telecom workers saw as ultimately destroying their technical skills and jobs. Bitter confrontation led to a form of settlement with Telecom management, but the dispute led the Fraser Government to establish a national inquiry into the wider effects of technological change, known as the Myers committee.[10] However, the rift between Telecom management and staff, and the Fraser Government, was deep, and ministers missed no opportunity to make life as tough as possible for all members of the statutory corporation. Interest rates on pre-1975 loans were changed, Telecom was banned from the overseas-loan market and prevented from using leverage leasing, and an additional $120 million had to be found by Telecom which virtually amounted to paying government an unanticipated dividend. Sir Philip Lynch's Razor Gang in 1981 demanded that 2000 workers be lopped from Telecom's workforce. Minister Staley used the $500,000 limit, on Telecom contracts which needed ministerial approval, to block its move into the marketing of small PABXs. Minister Sinclair prohibited Telecom from setting up a national videotex system in 1981, a move widely condemned not only within Telecom but also within the private sector.

Associated with these decisions was a broader intention to change the structure of Telecom to provide greater opportunities for the natural constituency of the Liberal Party, members of the business community. Ways had to be found to open up Telecom to the private sector. The notion of the domestic-satellite system as an alternative communications network to Telecom became an attractive option to the Fraser Cabinet. No one, however, in the Fraser Government, or in the communications-department bureaucracy, could grasp the detailed policy implications of a second national communications carrier.

After Davidson

Contemporary economists have examined the issue of why the telecommunications industry has been so widely dominated by public-monopoly institutions and why these organisations have enjoyed for so long the legal protection of their monopoly status.[11] Public-monopoly provision of national telecommunications has historically been defended on the grounds of 'natural' monopoly.

Basically, the argument for 'naturalness' is that economies of scale are such that it would be inefficient to have more than one firm producing the industry's output. The establishment of a public monopoly in the telecommunications field also assumed that, if a telephone network was developed by different companies, then the diversity of equipment used could mean that consumers might not be able to interconnect between the different systems. However, with one public organisation given sole responsibility for the national telephone network, compatibility of equipment would ensure that consumers could interconnect with each other, irrespective of their supplier or location.

Social equity factors are also involved, in that judgement must be made about those telecommunications services which are inherently uneconomic, such as telephones in remote country areas. Privately owned networks can hardly be expected to offer uneconomic services. Hence, public monopolies have been charged with the responsibility of the development of a range of national telecommunications services, and transfer income from services in densely populated urban regions to cross-subsidise uneconomic services in remote areas. Any moves to change the structure of a public monopoly in telecommunications ought to take into account the social objectives of the community as a whole.

Several economists who have specialised in this field argue against the notion of 'naturalness' of a communications monopoly and point out that monopoly has been legislated in order to distribute income rather than to enhance efficient resource-allocation. Surely, they argue, proof of the 'naturalness' of a monopoly ought to be found in the emergence of a single form of production without reliance on legislation. Now that many telephone networks around the world are approaching the position of universal access to a telephone, they ask, why have governments opted to retain the monopolies. Moreover, argues this school of thought, the economic barriers to entry created by legally protected monopolies have lowered the potential benefits to be realised from competitive action.

The privatisation winds of change were reflected in Australia by the Fraser Government appointment of a committee of inquiry into telecommunications services in September 1981, headed by Jim Davidson. The terms of reference of the Committee were:

Having regard to the continuing needs to provide adequate telecommunications services throughout Australia as efficiently and economically as possible and the significant technological advances which are now occurring in the telecommunications field both in Australia and overseas, the Committee is requested to examine and report to the Minister for Communications on:

a. the extent to which the private sector could be more widely involved in the provision of existing or proposed telecommunications services in Australia either alone, in competition with or in conjunction with the Australian Telecommunications Commission;

b. what consequential changes may be necessary in the statutory functions, duties, financial objectives and monopoly provisions of the Commission; and

c. the effectiveness of the Commission's operational policies and organisational arrangements.[12]

The prime assumption in the terms of reference was that new forms of competition in Australia's telecommunication industry would be beneficial. In essence the committee was expected to find ways to revamp Telecom to enable a select group within the Australian community, the private sector, a larger slice of the telecommunications action. For this task, the government well chose its members for the committee; Jim Davidson, Chairman of Commonwealth Industrial Gases; Malcolm King, former Deputy Manager of CSR Ltd; W. Dick, Chairman of Pacific Carpets; and Professor Tony Karbowiack, Head of Electrical Engineering at the University of NSW. Members of the committee, without exception, were philosophically well disposed towards the development of a blueprint for great privatisation of the Australian telecommunications industry. With a small secretariat and research staff, they were assigned by government not a broadly based investigation of telecommunications in Australian society, but a specific brief to fundamentally alter the role of the common carrier. The committee was not asked to work within a social framework, nor did it have any members who could explore the issues of how changes within the Australian community might be related to the processes of communication by a telecommunications statutory corporation. The Commonwealth government had established a national inquiry into telecommunications with narrow terms of reference which allowed the committee to virtually ignore social policy.

The submissions by various groups to the Davidson Inquiry encapsulated arguments which are ongoing in Australia, as well as many developed countries, concerning the most desirable model for national telecommunications, and the uneasy alliance of public and private-sector involvements in an information economy. Telecom management naturally approached the inquiry with trepidation, seeing it as the culmination of their confrontation with the Fraser Government. Their strategy was essentially to stand their ground and stress the achievements of Telecom since its formation as a statutory corporation in 1975, but to be accommodating to Davidson towards the private sector in specified areas of development. It was odd after such nasty industrial confrontations to see Telecom's workforce taking a stronger pro-Telecom line than Telecom management itself, but the ATEA knew that the stakes were high for all those in Telecom.

In its submission to Davidson, Telecom argued that it should remain as a statutory body, with modifications 'to recognise the existence of competition and to provide appropriate management flexibility.'[13]

Management argued that Telecom should maintain its role as the exclusive national telecommunications common carrier and continue to operate a cross-subsidy for uneconomic services. Several of its proposals acknowledged greater accommodation of private-sector demands. For instance, Telecom proposed that terminals and terminal systems should be progressively opened to wider competition to expand the market for the benefit of customers. On the question of private-line and network interconnection with the public network, Telecom argued that what was needed was 'simplified arrangements within which private networks can be operated', but the whole issue of interconnection was to remain vexed and unresolved for some time to come.

The Telecom submission stressed that if there was to be major change in Australian telecommunications then the potential benefits for co-operative ventures between Telecom and private bodies ought to be recognised, and that Telecom should have freedom in the future to market nationwide telecommunications services in competition with the private sector. Telecom was giving notice that if there were to be new rules for the telecommunications game, then as the incumbent giant it would be a formidable competitor for any newcomers. Of the many private-sector submissions, Business Telecommunications Services best encapsulated the private-sector case for change. BTS proposed that the Australian telecommunications market be divided into Basic Services and Enhanced Services.[14] The Basic Services were to be provided on a monopoly basis within the Telecom framework: local telephone calls, national trunk calls and the national telegram service were identified in this context. Enhanced Services would be provided within a commercial, competitive environment open to private-sector participation: the supply, installation, connection and maintenance of all terminal equipment, including PABXs, were examples. BTS essentially proposed that Telecom would maintain the least profitable established telephone services, together with the highly profitable STD routes, but that there ought to be new opportunities for private-sector participation in the terminal equipment market and in new telecommunications services. Central to the argument was that Telecom's public common-carrier monopoly should be broken, and replaced with competitive carrier networks. In this situation Telecom Basic would be separated from a new commercial entity, Telecom Enhanced, with Telecom paying taxes 'to ensure fair competition'. The essence of the proposal was to leave Telecom Basic with its traditional telephone network, but to break Telecom's monopoly over leased lines, telex, the Austpac service, the digital data networks and private networks, all of which services would be redefined as 'enhanced' services. It was a proposal for radical change.

The arguments for retention of a public telecommunications monopoly, or against its removal, are amongst the most contentious contemporary economic debates in developed countries, and there are broad similarities in the kinds of arguments presented no matter where the issues are debated. These key arguments on major telecommunications issues are discussed below, with emphasis on the Davidson inquiry in its Australian context.

Cross-Subsidies and 'Cream Skimming'

Telecom argued that it ought to maintain its role as the exclusive national telecommunications common carrier on the grounds of its national social responsibility to make services as widely available as possible. It does so by operating a cross-subsidy from profitable to uneconomic services. Telecom argued that its cross-subsidy was $290 million (in 1982) for uneconomic services, with key loss-areas of country telephone services, the telegram service, public telephones and metropolitan domestic services. If these services had to be charged at rates to ensure normal profit margins, then some of them would become so expensive that they would have to be dropped. Hence, if services such as remote telephone extensions and the public telegraph service are to be retained on the grounds of national social interest, they must be subsidised by those services which are highly profitable, such as the Melbourne-to-Sydney STD telephone route, known as 'the golden boomerang'. If the notion of a universal service at an affordable cost is a desirable national goal, then there is a case for retention of a form of monopoly which protects profitable services for the cross-subsidy.

Existing telecommunications monopolies in most countries use pricing practices which in effect tax long-distance calls and business users so that they can hold down the prices charged for local calls, residential users and rural subscribers. These desirable redistribution objectives would be threatened by a new policy of harmful competition which allowed new private entrants who wished only to provide services on the highly profitable inter-capital city routes. This kind of selective private-enterprise practice is referred to as 'cream skimming'. The corollary of effective cream skimming is that it will threaten the present range of services offered by the public monopoly and lead to increased prices for telephone lines and local calls and longer delays in installation and maintenance.

The BTS response to these arguments was that Telecom's level of profit from its trunk routes was sufficient to adequately compensate its loss-making areas. Their proposal argued that Telecom should retain its monopoly over the trunk telephone service, and that Telecom Basic would continue to operate the bulk of Australia's national telecommunications service requirements, which assured Telecom of keeping about 90 per cent of its established revenue base. However, the unstated variable in this argument is to what extent the introduction of new competitive private carriers which offered voice services would in the long run undermine Telecom's secure financial base. Wide freedom for private telecommunications operators must eventually divert revenues which impair the extent of cross-subsidy.

Davidson highlighted some anomalies in the cross-subsidy argument. The committee pointed out that Telecom's present product-costing practices reflected the difficulties of allocating costs and revenue to a mix of traffic and products when consideration of national benefit by cross-subsidy was involved. In fact, they argued, uniform pricing led to the misallocation of funds in that the financial burden for support of socially desirable objectives was borne by selective groups of customers, instead of being spread over the whole community. They challenged Telecom's pricing premise that low telephone rentals are provided for low-income families. For example, argued Davidson, price differentiation between business and non-business telephone rentals existed; the effect was that wealthy families enjoy the same low-income subsidy as non-business users, while 'struggling businesses who provide employment pay discriminatory high telephone rentals'.[15] Conversely, low-income families of metropolitan Australia were subsidising rural Australians, including wealthy rural subscribers. The Davidson committee was critical that cross-subsidisation dominated Telecom's existing tariff arrangements: call charges cross-subsidised rental charges, trunk calls subsidised country revenue, metropolitan services cross-subsidised country services and customers close to exchanges cross-subsidised customers remote from exchanges.

In their final recommendations, the Davidson committee brought into question the fundamental principle of whether it was the task of Telecom to make arbitrary judgements about the priorities of cross-subsidisation, or whether it was a matter for political judgement by the elected representatives. This alternative could mean, of course, that Telecom's cross-subsidy would become a matter of political football, dependent on which political party was in office, and especially on the amount of clout the National Country Party could command. Davidson stressed that Telecom should shift from uniform pricing to cost-related pricing practices. Procedures ought to be developed, Davidson argued, to provide direct government subsidies to any class of telephone subscriber deemed worthy of government assistance. This was an extension of present practices where the Commonwealth government reimburses Telecom for foregone revenues to specified pensioner subscribers.

Davidson did not call for the abolition of the cross-subsidy, but argued that Telecom ought to be responsible for fixing its own standard rentals and call charges, and that the cross-subsidy be reduced to a level which Telecom could absorb while remaining competitive. Ministerial approval for increased tariffs had always been standard practice since Telecom's formation, but in attempting to devise a more competitive telecommunications environment, Davidson's view was that Telecom must be able to fix its own tariffs. Telecom's pricing practice, the committee argued, ought to essentially reflect costs to a substantial degree and minimise price discrimination.

The costs of telephone calls with new pricing strategies became a contentious issue after the report was released. Davidson felt that extravagant allegations of price increases had been made about his proposed change. In a long 'letter to the editor' which appeared in the national presses, Davidson wrote that 'at present the telephone services of some 10 per cent of subscribers . . . are subsidised by the remaining 90 per cent because the price of telephone rental is the same regardless of location. If Telecom's prices are restructured as recommended by the committee, 90 per cent of subscribers will pay less and 10 per cent will pay more.'[16] The recommendation in the report for timed local calls, which Telecom estimated would cost $150 million to implement, got a great deal of attention in media coverage of the Davidson report.

Clearly, in the aftermath of Davidson, Telecom management has moved towards greater application of the user-pay principle in its pricing policies. Progressively, Telecom has adjusted its prices more in line with costs, and telecommunications users can expect continuing increased charges in specified areas. Telecom's charging orientation will be directed towards the anomalous disparity between trunk and local tariffs for all telecommunications services. It will depend on the future political climate as to the extent to which Telecom will offer a range of useful but uneconomic services.

Private, Independent and Leased Networks

An important component of the privatisation of communications is the development of new telecommunications networks. A network is a telecommunications system used for the transmission of electromagnetic signals by wires, conductors, cables, fibre optics, radio waves or other means. Telecom's national terrestrial 'core' network consists of conduit cables, aerial distribution and microwave radio plant which forms the supporting infra-structure for specific kinds of transmission facilities. Telecom has four dedicated public switched transmission systems in the national terrestrial network – the telephony, telegraphic (public telegram), telex and public data switched networks. The largest of these public switched networks is dedicated to the carriage of public telephone voice traffic; this network interconnects over 5 million telephone subcribers who use some 7.7 million handsets.[17] Apart from these four networks, the national terrestrial system has the capacity to provide other specialised transmission systems, such as private, independent or leased networks, which are developed for a variety of reasons.

A private network is one owned and operated by an organisation for its own use, and is not interconnected with the public network. For example, a mining company may have a need for an extensive communications network, but because of its isolated location, the PTT may not be able to justify the investment of capital in such a venture. In this circumstance, if private capital is available, then there is obviously a good case for the authorisation of a private or independent network to be established by the mining company. An independent network is a sub-set of private networks, owned and operated by an independent agency. Examples of private networks authorised under the *Telecommunications Act* 1975 include state public utilities (transport and energy), pipeline authorities, remote mining establishments and media-production houses operating closed-circuit television systems; examples of private networks authorised under the *Wireless Telegraphy Act* 1905 include Royal Flying Doctor Services, Qantas, police departments, radio paging-system operators, and taxi and courier-service operators.[18] A leased network is also used for private purposes, but the user leases lines from the common carrier. Examples of leased lines are television relay lines, tie lines where a company can lease capacity for its own internal telephone system, and data private lines, such as those used by banks to transfer data between branches or offices.

Each of these kinds of networks – private, independent or leased – allows organisations to design and operate a network to meet specific telecommunications needs. The problem for PTTs in general, of course, is that they invariably see dangers of private competition with monopoly carrier networks, or the potential to direct traffic away from the public-service authority. In its submission to Davidson, Telecom argued that, in providing 44,000 data services, 400,000 leased lines and 150,000 private lines, it satisfied the demand for private-sector networking.[19]

The US experience of deregulation of telecommunications is relevant in this context. Briefly, a benchmark decision by the FCC in 1969 allowed the first competition to AT&T in the long-distance transmission and switching market by allowing MCI Communications to build a microwave radio-relay network between Chicago and St Louis. After years of losses, MCI today has a network which covers the continent, incorporating MCI and their leased facilities, and is now the operator of the largest long-distance telecommunications network in the USA after AT&T. Private networks have proliferated in recent years in the USA, the most competitive private-telecommunications market in the world. Is this the best direction for the future of telecommunications, and should Australia allow greater extension and freedom for private lines? What is the likely revenue impact of permitted resale of leased Telecom lines? Would open resale of leased Telecom capacity, to a third party, encourage private exploitation of the national terrestrial network? How can anyone distinguish between leasees wishing to resell genuine surplus leased capacity from those who really wish to lease bulk capacity in order to compete with Telecom?

Davidson made several major recommendations on the resale of leased Telecom capacity and on the establishment of independent networks. Unrestricted use and resale of Telecom capacity should be permitted, according to Davidson, and prices of leased lines should be non-discriminatory. As well, independent networks should be permitted subject to ministerial authorisation.[20] Davidson then moved for a much greater accommodation of private-sector demands, and argued that with revised pricing policies Telecom would be able to protect its commercial base.

Interconnection

Interconnection of private networks with the national public network is technically complex and perceived by the PTTs as potentially threatening to their operation. In the public telephone network every subscriber is connected to every other subscriber, through the common-carrier network. Private networks are available exclusively to their own users, and may or may not use the common-carrier terrestrial network. Telecom has developed strict policies wherever private networks interconnect with the public common-carrier network. Naturally, a monopoly carrier is only keen to develop private networks when it clearly benefits from their introduction, and monopoly carriers tend to restrict access by competitive carriers who wish to use the public network in ways that are against their competitive interests. In recent years Telecom has felt obliged to move towards a more flexible policy on interconnection: its submission to Davidson acknowledged a need for simplified arrangements of interconnection between private networks and the common-carrier network, but Telecom did not elaborate on acceptable mechanisms to achieve greater interconnection flexibility. Private-sector protagonists argue that new technologies allow full scope for the intercommunication of services and that private-network operators, leased or independent, who interconnect with the public-sector switched network, can offer better services to their customers. And surely, they add, the business sector is entitled to the best range of telecommunications services.

The Davidson Report acknowledged that competitive leased networks based on Telecom facilities could lead to 'cream-skimming' attempts by private lessees, but added that Telecom could defend itself against such competition because the costs of leased line charges are established by Telecom. Davidson also came out in favour of unrestricted interconnection of leased Telecom capacity, with the recommendation that 'interconnection should be permitted between leased Telecom networks and national terrestrial public switched networks.'[21] Their position on interconnection of independent networks was more qualified: 'interconnection should be permitted between independent networks and the national terrestrial public switched network on the same basis as leased networks, provided that ministerial authorisation has been given for interconnection'.[22]

Responses to these sets of recommendations fell into two broad camps. There was a view which argued for greater private-sector flexibility for leased and independent networks and pointed out that Telecom was still empowered to establish interconnection charges for any private carriers who wished to make use of the public network. On the other hand, it was claimed that operators of independent and leased networks would progressively carry more telecommunications traffic, that competitors could pitch tariffs below Telecom charges, and that significant siphoning of Telecom's traffic would inevitably affect its overall profitability and threaten the continuation of the cross-subsidy.

Terminal Equipment and Services

Telecom authorises the attachment of terminal equipment subject to specific technical and electrical safety requirements. The pattern of market sharing between Telecom and the private sector for telecommunications equipment is complex and Telecom excludes the private sector from specified attachments. Telecom preserves its monopoly over the 'first telephone' and it has an exclusive monopoly for intercom, PABX systems, 'Commander' and Multicom systems. However, Telecom has permitted the attachment of privately owned interface devices which enable computers and communicating word processors to be connected in lieu of standard telex machines, and it has also chosen not to engage in marketing data equipment. The general private-sector view was that Telecom's existing supply arrangements, based on self-interest, inhibited the introduction of new equipment and reduced consumer choice. BTS argued that market forces ought to play a more significant role and that the private sector be encouraged to invest more extensively in the terminal equipment market.

Telecom management was receptive to the idea of greater competition in the area of network terminal-attachment equipment. They saw that opening up the installation and maintenance of terminal equipment would undermine the industrial strength of the technicians' union, and that open competition would allow Telecom to become involved in areas from which they had been banned by ministerial directive, such as the under-50-line private PABX market. Understandably, Davidson recommended that the private sector should be permitted to participate in all aspects of terminal-equipment marketing, and in the wiring of customers' premises as well.[23] What the committee failed to come to terms with was the key role that local manufacturing, especially Australian owned companies, might play in the future development of the system.

Standards and Network Incompatibility

With common technical regulations a PTT monopoly can guarantee user-interconnection for its equipment and services; the difficulties of ensuring compatibility of equipment are greater where there are multiple competing carriers. The development of telex in the USA provides a useful example of the disadvantages which may arise with competitive systems. A variety of different competitive services - AT&T, teletypewriter exchange services, TWX and Western Union's telex services - clearly disadvantaged the development of telex in the USA. The unregulated communications-equipment industries - computers, videocassette recorder, videodisc - provides a plethora of examples of disadvantages for consumers when intensely competitive environments prevent the interchange of software for different brandname equipment. Multiple competing international carriers in the USA have led to greater expense in network interconnection, in some situations, and greater customer inconvenience. If, for instance, a European PTT wishes to interconnect its data network with the USA, it first must deal with the various international licensed carriers and then with the main US domestic carriers to Telenet and Tymenet in a process which involves extra interconnection-equipment costs.

In Australia, the telecommunications network and terminal-equipment standards have essentially been controlled by Telecom, though several other public bodies are involved with telecommunications regulatory standards, such as OTC, Aussat, and the Department of Communications. In its submission to Davidson, Telecom indicated that it was prepared to relinquish its responsibility under Section 13 of the *Telecommunications Act* 1975 in favour of the development of a joint industry-Telecom equipment standards approval policy. BTS argued that a PTT can use its power as an equipment-regulator to improve its commercial position and that all regulation ought to be removed from Telecom. Approvals for attachments to the network, according to BTS, should be taken over by the Department of Communications who ought to sub-contract to the private sector wherever possible. Davidson recommended that responsibility for the establishment of standards should be transferred from Telecom to the Standards Association of Australia (SAA). The area of technical standards is administratively messy and long delays are inevitable in the preparation and acceptance of new rules by an independent authority - Telecom is likely to maintain its commanding position in the maintenance of adequate technical standards in the Australian telecommunications-equipment market.

How then can the proposals of the Davidson committee be

assessed in the context of the movement towards the greater privatisation of telecommunications in Australia? The committee essentially produced a pragmatic report rather than a radical set of proposals for change: their blueprint was neither a version of the British model to sell off bits of British Telecom, nor was it advocating a wholesale American deregulatory model. Primarily Davidson argued for Telecom to develop a stronger commercial orientation as a publicly owned commercial business and for the creation of new opportunities for the private sector, especially in separate networks and in the terminal-equipment markets. The report called for the establishment of an incorporated company – Telecom Australia Limited – to be owned 100 per cent by the Commonwealth Government. The Company:

should have all necessary powers to provide telecommunications systems in Australia in accordance with relevant statutory authorisations;
may enter joint ventures or partnerships as well as form or participate in the formation of subsidiary companies which have as an object the operation of telecommunications systems;
should place with its board of directors full responsibility for management of the business of the company. (R92)[25]

Telecom under this proposal would have to pay all government taxes, rates and charges, but in return would be cut adrift from ministerially imposed constraints such as on its loan-raising activities, financing methods and staffing levels. Davidson's chief recommendations represented a compromise, allowing greater competition for Telecom and the private sector, yet at the same time attempting to preserve the essence of the common carrier so that the public telecommunications network could be upgraded and developed.

The Davidson report did not call for open competition between Telecom and the domestic-satellite system. Had the Davidson committee accepted the notion of full deregulation, they could have recommended that Aussat, an independent network with a national telecommunications infra-structure, be allowed to fully compete with Telecom. They acknowledged that the establishment of Aussat would introduce competition with Telecom for some long-distance traffic, but argued that this would not damage Telecom, provided the terrestrial carrier restructured its prices. Aussat should be permitted to operate leased terrestrial networks, but not allowed to own local terrestrial networks. Aussat, according to Davidson, ought to adopt a constructive trading relationship complementary to Telecom.

In many ways the Davidson committee gave the Fraser Government the kind of report they appeared to want. Soon after its release, Neil Brown, then Minister for Communications, praised the report for providing 'an excellent charter for the development of communications in the future' and said that 'it reflected world wide trends in telecommunications policy.'[26] The Minister added that its findings would be carefully considered before any decisions were undertaken. Members of the National Country Party undoubtedly noticed the wide publicity given to the possibility of substantially increased phone charges, especially for rural areas. Within a few months of the release of the report, the Liberal-National Coalition government had again called for an early election, in March 1983. During the campaign period, Brown issued a statement 'to allay community concern over some of the recommended changes.'[27] Brown said that though the Government had not concluded its considerations of the Davidson Report, the central recommendation was unacceptable – that 'Telecom's monopoly of basic telephone and telex services should be broken and the field opened to private enterprise.' To this confusion he added a vague rider that 'the Government's decision still left ample room for private enterprise to compete in the provision of new telecommunications technology.'[28]

The Davidson exercise was yet another example of the muddled and indecisive way the Fraser Government approached the vexed issues of communications policy. As a casestudy in policy formulation the chief stages typified the Fraser administration's approach to forward strategies – the notion of greater privatisation, followed by a government report which showed how this might be achievable, then internal tensions and inability to cope with complex matters which led to a break-down. A further complication in this case was that another government inquiry, which ran simultaneously with Davidson's, the Australian Broadcasting Tribunal's inquiry into cable and subscription television, produced some recommendations concerning Telecom's possible role in cable-television innovations which were contradictory to the general telecommunications inquiry. The overall result for both of these inquiries was that decisions were shelved and the status quo was maintained.

The ALP had a different set of agonies in coping with the Davidson Report. In 1982, shadow Minister for Communications, Senator John Button, presented a submission to Davidson which astonished some ALP members, particularly Telecom employees, by its acceptance of the concept of greater competition for the common carrier. The pressures on Button to change his stance and support the retention of a publicly-owned single telecommunications carrier have been described elsewhere.[29] By the time the Davidson Report was released, Button was publicly arguing that Telecom 'needs a complete network over the whole country to work effectively and the alternative is a fragmented, chaotic telephone service.'[30] Leading Labor protagonists produced their interpretation of the Davidson Report in terms which were electoral dynamite. According to Bill Mansfield of the ATEA 'ordinary telephone accounts would rise by more than $100 per year.'[31] Mansfield, skilled media operator, argued that three of Davidson's recommendations would inevitably force up subscriber costs: that Telecom be subject to government rates and taxes, time-based billing for local calls and the dismantling of cross-subsidisation. The ATEA's scare campaign on future Telecom pricing was highly effective – what counts most of all in policy formulation in Australia is what can be thrown up on the hustings.

The Hawke government rejected the main thrust of the Davidson Inquiry and officially endorsed Telecom's monopoly as the national telecommunications common carrier in October 1983. Minister Duffy, using ATEA rhetoric, announced that 'this government's policy is to ensure that the best possible services are available to the Australian public, and that universal access is provided to the national telecommunications system at a price all can afford.'[32]

Within less than a year of Duffy's statement, Opposition Leader, Andrew Peacock, in announcing the coalition's shadow economic-policy blueprint, called for greater privatisation of Australian industry, and indicated that Telecom's status would again be re-examined after a change of government.[33] *Déjà vu!*

In the marketplace, strong support remains for greater privatisation. Though BTS has disbanded as a unified group, many of its former members remain active in the politicisation of issues in telecommunications. PBL, for instance, in August 1984, boldly challenged the authority of the Commonwealth Government in the High Court to control radio transmissions. Summonses issued against Channels Nine, Sydney and Brisbane, alleged that between December 1981 and June 1983 they maintained, without authorisation, a station for the purpose of transmitting or receiving messages by means of wireless telegraphy, contrary to the provision of Section 6 (1) of the Wireless Telegraphy Act.[34] The allegation was that Channel Nine had created and maintained stations at Somersby, NSW and near Coolangatta, Queensland, to serve as a link in a chain of private links as an alternative to the Telecom system. Channel Nine mounted their defence based on Section 92 of the Constitution which provides for freedom of trade, commerce and intercourse among the States. This is unresolved at the time of writing, but the case demonstrated the degree of commitment on the part of a major Australian television network to opening up private networks in opposition to Telecom.

BTS's role appears to have been assumed by ATUG – the Australian Telecommunications Users Group – with more than 180 members including most major banks, airlines, mining corporations and insurance companies. Its first full-time director, Wally Rothwell, appointed in mid-1984, described ATUG as 'a powerful lobby group representing some powerful members.' ATUG has called for 'a relaxation on telecommunications equipment and maintenance and greater access to worldwide telecommunications technology.'[35] Like BTS, ATUG, under a veil of professional seminars, is strongly committed to an attack on Telecom's interconnection policy and it forms a collective bargaining group in dealing with Aussat. It is particularly committed to getting greater freedom of choice in the selection, interconnection, operation and maintenance of PABX systems and is highly critical of the mandatory requirement that Telecom perform the maintenance of equipment connected to PABX systems. ATUG is likely to become particularly active in the area of the corporate use of Australia's domestic-satellite system. A change of government would undoubtedly result in more intense lobbying for greater privatisation and a higher public profile for ATUG.

So whither Telecom? Australia's biggest public corporation is placed in an invidious position in that it operates in a country where the two major political parties, partly because of their different power bases, have divergent perceptions of the future of telecommunications. The ALP, with its strong union base, supports the retention of the common-carrier public monopoly, while the Liberal Party pushes the business-services line of greater privatisation, and toys with dismantling the monopoly. In this political climate, and especially because of the volatility of the Australian political system, Telecom management has opted to be seen to have a bit each way with both major political parties. During the period of the Davidson investigation, Telecom management showed considerable political astuteness by its avoidance of provocative public debate and by showing its intentions of developing closer relations with the private sector. Yet, after the return of the ALP to government in 1983, the traditional arguments in favour of public telecommunications monopolies resurged.

Davidson's work will certainly have long-term effects on Telecom. There are some general indications of movement within Telecom towards more direct cost-related pricing, and a shift in tariff priorities towards the business sector at the expense of private householders.

These trends have been monitored by the Australian Federation of Consumer Organisations, who claim that questions of dependability of service, efficiency of supply of services and adequacy of complaints procedures, should be considered in the broader context of public participation in Telecom. This organisation has suggested that consideration be given to the appointment of a Telecom Ombudsman to oversee the existing dispute-resolution procedures, or the appointment of a consumer commissioner to the Board of Telecom, or the establishment of a Telecom Users Association subscribed to by Telecom users at minimum cost.[36] What ought to be remembered is that, unlike the ALP or the Liberal Party, Telecom's constituency is the Australian community as a whole.

8 The Domestic Satellite: Australia Joins the Club

In October 1979 the Minister of Postal and Telecommunications, Tony Staley, announced to the Parliament that Australia was to have its own domestic satellite system. We were told that 'it is all too easy to overlook, or remain blissfully oblivious to, the plight of those of our fellow countrymen who are seriously disadvantaged by a lack of communication services and communication-dependent services.'[1] A series of 'sobering facts' were presented to show the inequality of telecommunications and broadcasting services – that 40,000 people in isolated communities had no prospect of being connected to a telephone in the foreseeable future, that 500,000 people in Australia had no television service at all, and another 220,000 people had an ABC service only, and that nearly four million Australians had only one national and one commercial television service. The situation with radio, it was argued, was little better – 375,000 people lived beyond the normal reach of ABC stations and some four million Australians received only the third ABC network.[2]

An argument for the introduction of a new technology based on equity of communication services obviously has merit. Undoubtedly domestic satellites have great potential in attempting to overcome communication disadvantages for Australians who live in isolated areas. Surely the people of Oodnadatta, Bamaga or Burrow Creek deserve comparable television services to those available to fellow Australians who live in Melbourne or Sydney.

Five years later, and about a year before the launch of Australia's first domestic satellite, a letter to the editor appeared in the *Australian* (29-30 September 1984) from Eric Michaels on the isolated Yuendumu Community in the Northern Territory. It read in part:

The fact is that most of the people in Australia who have no television at all are Aboriginal people. And there is reason to question whether the kinds of mass media being discussed are appropriate or desirable from an Aboriginal perspective.

But the critical fact is that most of these people also have no automatic telephone service...

It is not just a matter of personal convenience; the cost and inconvenience of supply and maintaining vital services are outrageous.

The ability to plan, co-ordinate or simply be informed is minimal. The feeling of isolation is complete. How ironic that soon we will be able to choose between perhaps several channels of urban television, but be unable to order a critical part for the power plant, co-ordinate educational services, or discover when a close relative has died. Make no mistake, telephones are the top priority for what AUSSAT could bring us, as the Australian Broadcasting Tribunal recently reported. But the Government isn't listening...

Lets simply discard the ridiculous premise that the new satellite has anything to do with servicing the outback. If that were the case, very different issues would be under discussion and remote Aboriginal people would be directly involved.

Tony Staley's and Eric Michaels' statements encapsulate the wide divergences of opinion about the justification for a domestic-satellite system, and what needs it might serve. As an exercise in communications policy, the introduction of the domestic-satellite system shows the monumental failure of the Australian political system to come to terms with complex issues related to technological change. The then Minister, and the Department of Postal and Telecommunication (now Communications), must shoulder a great deal of the blame for not establishing an adequate public investigative process out of which could emerge clear-sighted satellite policies. The series of satellite inquiries, later to be examined, which led up to the ministerial announcement, failed to come to grips with the fundamental questions about an Australian satellite system – what communication needs were presently inadequately serviced, how we might act on those needs and which technology was most suited, what the associated community benefits and/or problems were, what costs were involved and how new national technological projects might be properly managed.

Instead, the wheels of policy making essentially fumbled along, responding to pressure points and reacting to problems.

An Alternative or Second Telecommunications Network?

The driving force behind the proposal to introduce an Australian domestic satellite came in 1977 from a commercial broadcasting organisation, the Channel Nine network controlled by Kerry Packer. Government acted quickly on Packer's idea and accepted the recommendations of government inquiries dominated by technocrats and bureaucrats. By accepting the technical and managerial expertise sifted by satellite inquiries, government effectively took the debate out of the public realm. Packer's concept, though modified by a process of objection, essentially became Government policy in 1979. Successive ministers and a new Government have subsequently grappled with conflicting claims from a variety of vested interests and agonised over finding ways of making the project politically acceptable and economically feasible. For several years it seemed Australia's domestic satellite project, initiated by the Fraser Government and endorsed by the Hawke administration in 1983, was desperately in search of a *raison d'être.*

Two polarised discourses underpinned the political decision to proceed – a school of technological determinism, as opposed to several random groups who might be collectively labelled as a school of perceived threats. The prime assumption of the former school was that Australia would get 'far behind' if we did not follow the rest of the world into the satellite age. For the second group, Australia's domestic-satellite system was a prime example of how the introduction of a new technology could create profound change to the existing order. Protagonists in both camps often demonstrated their transparent vested interests.

There were plenty of technocrats who pointed to satellite developments in the rest of the world as justification for Australia to follow suit. Canada and the USA had their own domestic satellites in the mid-1970s, our neighbour Indonesia has had a domestic satellite operating since 1976 and less developed countries such as Mexico and Brazil entered the 1980s with contracts and satellite-launch commitments. The question of whether we would join the domestic-communications space club was never really an open one. Whilst Australia has been comparatively slow to embrace this new technology, the forces of technological determinism assured its domestic existence and we are committed to the launch of two domestic satellites in 1985 and a third between 1986 and 1988. The major question since 1979 has been how we can ensure that the national satellite project operates to the best advantage of the Australian community.

Since it is a public venture, with establishment costs paid for at a time of record national public debt, what are the long-term cost benefits to the Australian community? Will it serve previously neglected communication needs, and lead to a new era of valuable information services? Might the project produce serious disruptive effects on the established telecommunications and broadcasting industries? Will it encourage a range of new business opportunities for hardware and software products? If so, will these be produced by indigenous Australian companies? To what extent will it compete with Telecom, which was established as a public monopoly with exclusive telecommunications rights? To what extent will the satellite authority milk revenues from Telecom for telephony, broadcast and data transmission? Will the satellite system directly compete with, or merely complement, the range of services operated by the terrestrial authority? Why, asked Telecom unionists, embark on innovation that might jeopardise the stability of Australia's most financially successful public corporation?

In the late 1970s when the Fraser Government was committed to the rhetoric of monetarism and to the greater privatisation of commerce and industry, it saw Telecom as an intransigent, badly managed public authority with strike-prone unionists. Little wonder that, when a major new technology was emerging overseas that offered the prospect of fundamental changes to the telecommunications system, it fell upon receptive hearts and minds in government. At that time in the USA there was a series of major corporate domestic satellite ventures. IBM, for instance, received approval from the FCC in 1977 to develop their successful subsidiary company, Satellite Business Systems (SBS). IBM's Australian managing director, Allan Moyes, captured the mood for change with his observation that 'monopoly communication carriers with a vested interest in existing facilities and an obligation to provide nationwide telephone communications appear far less motivated than entrepreneurial entities to bring these applications into commercial operations.'[3] IBM, as we have seen, was a member of the professional lobby group, Business Telecommunications Service, formed in the late 1970s to give effective political clout to the dismantling of the traditional telecommunications monopoly in Australia. Clearly, the sub-text of the government satellite initiative in 1979 was the development of an alternative telecommunication network to Telecom. This was the crux of government satellite strategy – it was not merely the technology to provide services to outback Australia.

Ministers involved with the satellite under the Fraser administration were cautious in their public statements about the prospects of offering services alternative to those of Telecom. Early in the debate Minister Staley referred to the 'flexibility' advantage: 'strategic considerations also feature in some minds: the concept of systems diversity where there will be in the future, with the addition of satellite domestic communications, less reliance on traditional systems and therefore more flexibility in times of emergencies. . .'[4]

Later, Minister Sinclair in a major address to the Parliament in 1982 went further – 'the satellite system has the potential capacity for competitive telecommunications networks and alternatives to the existing terrestrial communications system.'[5] Precisely what this meant in terms of which services were to be offered by whom, and how, was not made clear. Indeed, for the entire period of the Fraser administration, the relationship between Telecom and Aussat, the satellite authority, was never spelled out.

Whilst Telecom was confused and threatened by the prospect of the introduction of a domestic satellite, Aussat was never sure of how opportunistic it should be in public. Some fundamental contradictions emerged in Aussat's public rhetoric about the extent of its role as an alternative telecommunications authority. The general manager of Aussat, Graham Gosewinckel, told a commercial-television industry seminar in September 1982, that Aussat would usher in a new era of diversification for Australian users, freeing them of dependence on Telecom. He then argued that many had overlooked the significance of the Government's announcement earlier in the year that individual satellite users would be able to own their own earth stations, either to receive or to transmit and receive. He described that decision as 'the situation which broke the Telecom monopoly.'[6] He was speaking, of course, at a time when government telecommunications rhetoric was deregulatory, and with the results of the Davidson inquiry pending. Nine months later, with the election of a Labor Government firmly committed to the retention of Telecom's common-carrier status, Gosewinckel was saying the opposite. At a time (mid-1983) when the ATEA was running a well concerted campaign to scrap the satellite, Gosewinckel went public for Aussat, stating on ABC radio that 'the satellite will complement the terrestrial network', and later that 'the satellite should not be seen as some sort of competitor to Telecom.'[7] Such is the need for senior staff members of statutory corporations to 'read' the political climate of the day!

Another area of perceived threat was that the introduction of a domestic satellite system would contribute to the centralisation of information-flow in service industries. The satellites, the argument goes, with their vast capacity for efficient data transmission, will help speed the processes already underway where workers in the branch offices are progressively replaced by remote computer terminals, automatic bank tellers and by word processors which can communicate with each other. So the central office, with its central computer, replaces antiquated methods of information transmission, including decisions made on the spot by people with a knowledge of local circumstances. For some bank workers, satellites are seen as yet another example of new technology replacing the need for people. Also, in some broadcasting organisations the introduction of a domestic satellite system was seen to create instability within the sections of the television industry, a fear most commonly voiced by regional television operators who felt threatened that they would be replaced by television relay stations operated throughout the entire continent from a central distribution centre, probably located in Sydney.

If the introduction of an Australian domestic satellite system brings profound change, or contributes towards unforeseen and undesirable effects, several old alarmists will be quick to point out that their warnings went unheeded.

Communication Satellites

Australia, with its population heavily concentrated in scattered seaboard cities, and with small sparsely populated regions separated by great distances, has the kind of geographical communication problems that satellites were designed to solve. An orbiting satellite can transmit signals to any point on the continent where a suitable receiving dish on the ground can pick up the signals for relay to nearby places. A communications satellite is basically a conduit, or a method by which signals can be transmitted. Satellites themselves do not make any new forms of communication possible, but are merely delivery systems, for broadcast signals, telephony and data. A satellite is essentially a radio repeater, using radio frequencies, with similar antennae and radio equipment. It was physicist and science fiction writer Arthur C. Clarke who first theorised the engineering principles of satellites. In his 1945 four-page paper, 'The Space Station – Its Radio Applications' published in *Wireless World,* he postulated a chain of space stations in the earth's equatorial plane that would remain in the same spots in the sky from the point of view of terrestrial observers. The fundamental principle was to identify a corridor in space above the equator where a satellite would circle the earth at the same rate as the globe rotated. So communication satellites, to remain in the same position relative to any given point on the earth's surface, rotate once every 24 hours, the same period it takes the earth to rotate once on its axis. This orbital position, sometimes referred to as Clarke's orbit, of altitude 22,300 miles (36,000 km) above the earth's surface, is known as the geosynchronous or geostationary orbit.

There are two major components to a satellite system, the space segment and the ground segment. In the space segment a satellite consists of a number of microwave antennae to receive and transmit radio signals – radio receivers, translators to change the frequency, and transmitters – a package known as a transponder. Satellites generally have 12 to 24 transponders, each capable of carrying several hundred two-way telephone circuits, or one or two television channels per transponder. All the transponders share common antennas, usually one or two for receiving incoming signals – the uplink – and two or more for transmitting signals back to earth – the downlink. The satellite is powered by solar cells. To keep the antennas pointed towards the earth, some form of gyroscopic stabilisation must be imparted to the satellite to maintain its position in orbit. Spin stabilisation is now the usual method where the satellite resembles in silhouette an opened tin can which spins on its vertical axis whilst the 'lid' of the can, the antennas, are kept stationary with respect to the earth by an electric motor.

The second major component of a satellite system is the ground segment. Apparatus used to send signals to the satellite, or receive signals from it, are referred to as earth stations. They may be divided into two categories: transmit-and-receive earth stations or receive-only earth stations. They usually consist of dish shaped antenna and associated microwave receivers and transmitters. This technology is similar in principle to that used in terrestrial microwave systems, although satellite receivers are more sensitive than those used terrestrially.

The difference in size and cost of earth stations depends upon the trade-off between the design parameters of the space sector and the earth sector. The general rule is that the more powerful the signal from the satellite, the less sensitive and smaller the earth station; the weaker the signal from the satellite, then the larger, more sensitive and more expensive is the earth station required to pick it up. The strength of the signal can be increased in two ways, either by increasing the power of transmission from the satellite itself, or by concentrating the radio energy from the satellite into a narrower beam that is concentrated towards that part of the earth where it is required. This is like focusing a spotlight on a part of a surface to be lit, and the area 'lit', in the jargon of satellites, is a 'footprint'. This footprint can be designed to cover the entire continent or be concentrated on a smaller area, with an obvious correlation with the size of earth stations required. The satellite itself has a finite amount of power available. Hence if designers wish to offer the maximum number of channels within the limits of this power, then the output power of each signal must remain comparatively low. These are key principles which affect the design of a satellite system: the choice is to optimise a combination of satellite power, earth-station size and footprint size.

A satellite system offers several advantages over terrestrial-based communication links. Signals delivered via satellite suffer comparatively less atmospheric interference. Most radio-signal interference, meteorological or man-made, occurs in a narrow band of the atmosphere up to about 20 km above the surface of the earth. A radio signal sent from Sydney to Perth via terrestrial relays covers about 3500 km through the densest part of the atmosphere, whereas by satellite, though the signal covers about 72,000 km, only the first 20 km of the Sydney uplink and the last 20 km of the Perth downlink are within that part of the atmosphere with the greatest risk of interference.

Secondly, the cost of satellite transmission is independent of the distance covered. With terrestrial systems, using microwave relays, cables or open-wire lines, the signal distance travelled actually determines the cost. With a satellite, however, the cost of a phone call from Sydney to Newcastle ought to be the same as from Sydney to Perth, since the signal travels over the same distance, up to the satellite in space and back to earth. Moreover, the cost of transmission is inherently independent of the number of points to which the signal is directed. That is, one signal transmitted up to the satellite can be re-transmitted to any number of earth stations which can receive that signal anywhere within its coverage beam. This characteristic of point-to-multi-point distribution is why Indonesia, with its 300 islands, has adopted satellite technology for its domestic telecommunications. Satellites can remove distance from the cost formula so domestic satellites are only really cost-effective where long distances are involved and where there is guaranteed high demand. They are ideally suited to the point-to-multi-point delivery of national television networks.

Australia and International Satellites

An important distinction must be made between domestic-satellite systems, owned and controlled by nation states or private corporations, and those satellites of the international system run by the International Telecommunications Satellite Organisation, or Intelsat. Over the past twenty years this organisation has developed an integrated global satellite system to provide telecommunications services to scores of countries. It is through Intelsat that pictures of international news and sporting events so familiar to Australian television audiences are received – a Royal Wedding, an attempted Presidential assassination, or the sports perennials of Wimbledon, the World Cup final and the Olympic Games. The Intelsat IV A satellite over the Pacific Ocean covers the whole of Australia and carries programming mainly from the USA. The Indian Ocean Intelsat V satellite covers Australia west of Queensland, New South Wales and Victoria and mainly carries programming from European and UK sources. Australian demand for access to international television programs via the Intelsat system has shown staggering increases in recent years.[8]

The irony of such international togetherness is that the research and development costs for early satellites were derived from military sources, and launch vehicles were developed from intercontinental ballistic missiles. Without the massive investment of US funds into their military and space programs, satellites could never have reached their present-day level of sophistication. Intelsat's first satellite started in 1965 – a lone Early Bird satellite above the Atlantic Ocean to provide service (240 telephone circuits or one low-quality black-and-white television channel) between North America and Europe. In 1969 the initial global system was completed to bring television images of man's landing on the moon to an estimated 500 million viewers throughout the world. The American space program was the catalyst for phenomenal development of international satellite technology. The Intelsat organisation, formed in 1964 with thirteen members, has a current membership of 109 countries. The Intelsat system is comprised now of 16 communication satellites, deployed over three oceans (Atlantic, Pacific, Indian), with 658 earth stations in 157 countries.[9]

Australia was a foundation member of Intelsat. In 1962 the US Congress created a private company to co-operate with other nations in the establishment of a global communications satellite system – the US Communications Satellite Corporation, or COMSAT, which managed the system until 1976. The organisation of Intelsat is based on the principle of membership and voting according to investment quotas, or usage of the satellites within the system, vested in two assemblies and a Board of Governors. The USA is the largest user, earning it about a 23 per cent share of the system; Australia's ownership holding in Intelsat is 3.5 per cent, valued at $A23 million, which represents the fifth-largest shareholding in the system.[10] The initial attraction of joining Intelsat was for inter-continental telephony links, with international television links as a secondary motive in terms of prospective usage. Intelsat is commonly responsible for the ownership of satellites in space, but member countries own their own earth stations, which are in turn connected to their national telecommunications networks.

Australia's representation in Intelsat has been by members of OTC, the Overseas Telecommunications Commission. Within a year after its inception Intelsat had its first commercial communications satellite in operation above the Atlantic Ocean. Australia's first earth station, at Carnarvon in Western Australia in 1966, provided communications from Australia to the USA for NASA via a Pacific Ocean satellite as part of the American space program. The first live international satellite television link up to Australia took place in 1966, and the technology became accepted in Australia with the coverage of the 1968 Tokyo Olympics. By 1969 OTC had four earth stations in service, used for international telephone, telex and television services with the USA, Canada, New Zealand and Pacific Island countries, via Intelsat satellites. Impressive in the sophistication of its operation, yet restrictive in its range of sources from which television programming is derived for Australia, the system has never become McLuhan's 'global village'. We have seen all too few alternative programs to the main staple diet of commercial television delivered from other countries via satellite.

Intelsat, a rare successful model of international co-operation, seems no longer immune from the mounting pressures for the greater privatisation of telecommunications systems. *The Communications Satellite Act of 1962,* which outlines the principles upon which the US Congress founded Intelsat, has the fundamental aim of a single global telecommunications network. Now, however, four American based companies want to launch their own satellites to compete for trans-Atlantic business, Intelsat's golden boomerang. Two other companies want to carry traffic between North and South America, and Intelsat will also face stiffer competition when the eighth trans-Atlantic sea-bed cable between Europe and North America becomes operational in mid-1988. The establishment of separate heavy-route international telecommunications systems could endanger the viability of the Intelsat system. Intelsat's concern is that by allowing private satellite systems to serve only the best traffic routes, the economies of scale are drastically changed. Rate increases, to offset traffic losses, would then have to be borne by all members, including poorer member countries. In the wrangling with Intelsat and American officials, Australia's OTC has called upon the US government not to allow private satellites to operate in international competition with Intelsat.

As the demand for international television traffic has increased, Australian television networks have moved towards full-time leasing arrangements with Intelsat, as well as ownership of their own earth stations in Australia. In February 1983, Channel Nine became the first private organisation in the world to acquire a full-time lease of an Intelsat transponder, to enable it access to American satellite-delivered program packages in a five-year contract at an estimated cost to the Packer network of at least $6 million annually. This unique delivery system selects program material from incoming satellite sources within the US at Nine's Los Angeles complex, delivers the signals via microwave up the mountains to an earth station at Santa Paula, 80 km north of LA, where they are transmitted via Intelsat IV A to Channel Nine's own earth station at Willoughby. The company set a precedent by winning government approval in 1982 to construct and own an earth station in Australia. Normally Australia's satellite common carrier, the Overseas Telecommunications Commission, brought in transmission from the USA via its own earth station at Moree, for delivery via Telecom's bearers to a Sydney television studio. Four months later, ATN-7 in Sydney followed suit with the establishment of their own earth station at Epping, NSW. The delivery system is similar in principle to that of the Nine network – programs received at their LA complex from within the USA are transmitted via Intelsat IV A to the dual receiving dishes at Epping.

Private ownership by Australian television corporation of their own earth stations is a benchmark telecommunications-policy decision, because it undercuts the established common-carrier role of the OTC. The Seven and Nine networks are required to pay OTC, Intelsat's Australian member, for the downlink service, but it gives them greater freedom to bring in via satellite what they like, when they like, than does the conventional OTC pay-as-you-use satellite service. No OTC earth-station facilities at Moree and Paddington are used on this system and the cost of terrestrial links provided by Telecom are also eliminated. However, the networks remain customers, not common carriers – both Seven and Nine must use the program material for their own purposes and re-sale to another broadcaster, or sub-lease, would be in breach of their contract. OTC, of course, would have much preferred that this precedent had never been established, so that they could continue their monopoly on international satellite-delivered services within Australia.

The decision typified the deregulatory mood in broadcasting and telecommunications of the Fraser Government. Private ownership of earth stations has put in place the beginnings of an infrastructure which could be used to carry other traffic. It would be technically possible, for instance, to bring in data in the form of teletext with the television signals delivered from overseas and deliver the data elsewhere in Australia via uplink from their own earth stations. It would be technically possible to establish microwave routes between capital cities for a data-transmission service based on the capacity delivered to these privately owned earth stations. However, this is illegal at present under Section 6 of the *Wireless Telegraphy Act*, which bars the unauthorised establishment, erection, maintenance or use of 'any station or appliance for the purpose of transmitting or receiving messages by means of wireless telegraphy.' Such restraints, of course, can always be changed by the political process. Australian experience with international satellites changed the telecommunications parameters even before the launching of our domestic satellites.

Origins of Aussat

Some of the best entrepreneurs in Australia are to be found in Australia's media corporations. The present Australian domestic-satellite project grew out of the ideas of Kerry Packer's Channel Nine organisation in 1977. Packer has said that his staff had convinced him of the 'enormously exciting and fantastic' benefits that communication satellites had brought to the small hamlets with 25 or 30 people in Canada and Alaska – domestic satellites brought TV programs, radio programs, news, direct phone dialling, 'the mainstream of the nation, into every home, no matter how distant.'[11] Packer and his staff were well aware of many implications for the introduction of a domestic satellite to Australia, such as the possibility of the introduction of American style networking which would enable Channel Nine programs to be beamed to almost every part of the country, including his then sporting spectacular, World Series Cricket. Moreover, according to his senior executives, there would be the opportunity for the first time to bypass the Telecom terrestrial distribution system, at one-fifth of the cost.[12] Packer later explained the response of the then Prime Minister Fraser to his case for an Australian satellite based on rural needs:

> And I went and saw the Prime Minister and I explained to him my understanding of what was happening in those areas, and to his undying

credit he grasped on to it immediately and said, 'Of course it's what we want. It's exactly the sort of thing we need to stop the drift of people into urban areas. We can keep them informed. We can allow them to participate in whatever's happening around the nation.' And he said, 'Can you do more for me on this?', and I said, 'It will take six or nine months. We will pay for it, then we will put to you a paper of how we see this developing.'[13]

Kerry Packer commissioned American communications consultant, Donald S. Bond, who prepared a report on the desirability of an Australian domestic satellite, *The Opportunity for Television Program Distribution in Australia Using Earth Stations.*

The Bond Report argued the success story of satellites overseas and put up a concrete proposal for an Australian domestic satellite. Bond claimed that 'the satellite systems in North America have been successful beyond all expectations' and that 'Australia is not alone in striving to attain the goals of high quality colour television service throughout a large area comprising both regions of dense and very sparse population.'[14] His plan was essentially for a satellite service to distribute four television programs simultaneously to 90 communities throughout Australia. This was consistent, Bond argued, with government policy of a diversity of television programs available to all citizens. It is important to remember that it was the argument for enhanced conventional television services that dominated early debate on the satellite, though Bond added that other important services that would become available were instructional television such as health-care delivery by telemedicine, long-distance telephone trunks, closed-circuit television and digital data transmission. He advocated the use of a standard American design of television-networking satellites, operating in the 4/6 GHZ band, as 'this is the only practical choice available for the fixed satellite service in Australia' and 'the same band is being used with great success in North America in the Intelsat system.'[15] The overall cost was estimated then at $121 million for a satellite system that could begin operation within 30-36 months of the issuing of all authorisations and the placing of a contract for the spacecraft. The Bond Report concluded that the satellite might be the catalyst in 'bringing private enterprise and the Government together in a co-operative undertaking for the benefit of all the people of Australia.' His final word left little doubt that he strongly favoured a private-enterprise operation and that the resources and leadership of his client, Television Corporation Ltd (now PBL) were 'available to the nation'. The notion of a 'Packer bird' in space did not exactly get a good press from media arch-rivals, nor did it go unnoticed by OTC or Telecom management.

Within six weeks of the release of the Bond Report, the then Minister for Postal and Telecommunications, Eric Robinson, announced the establishment of a Commonwealth Government Task Force to inquire into all aspects, 'social, economic, technical, policy and other issues', related to a national communications satellite system for Australia. It was chaired by Harold White, General Manager of OTC – a senior staff member from a satellite authority – with senior public servants from the Departments of Postal and Telecommunications, Defence, Health, Finance, Transport, Telecom and OTC, an impressive array of technocrats and bureaucrats. Oddly, though television interests were the prime motivator for the introduction of a satellite, they were entirely unrepresented on the Task Force, with no one represented from commercial television, the ABC or the statutory broadcasting authority, the Australian Broadcasting Tribunal.

The report they ultimately produced dealt primarily with satellites overseas, potential domestic users, system models, ownership and control and legislative implications. Matters such as health, education, employment and the Aboriginals were dealt with in 13 pages of a 228-page report. Rather than beginning from a base that attempted to establish long-term effects, needs and benefits, the Task Force basically synthesised the information furnished by 159 contributions from interested parties. Little wonder that governments of the future were to flounder over this approach to economic and social investigation.

What emerged clearly from submissions to the White Task Force was the complexity of interests within the Australian television industry, and the lack of support for the Packer proposal. Publishing and Broadcasting Ltd suggested the establishment of three Australia-wide commercial networks for all Australians to receive the full range of program alternatives available in the major metropolitan networks, to be delivered via satellite. Such a proposal would require 200 new commercial television stations with about 150 associated translator stations, increasing the availability of commercial channels from one to three in the majority of regional areas, and from zero to three in some remote areas. A further 90 stations, to operate as independent or public broadcasters, were to be provided by additional high-powered terrestrial transmitters. It was implicit in the proposal for new commercial stations that the growth in advertising revenues by the mid-1980s, when the satellite system might be operational, would be sufficient to support both existing and proposed stations. Savings would accrue, it was argued, by reduction in the use of terrestrial circuits and through more efficient co-ordination and distribution of program and advertising material. If ever Australia had a chance to go down the deregulation road and opt for the survival of the fittest in the television industry, this was it.

The Packer plan was strongly opposed by most interests within the commercial-television industry. Regional-television operators were outraged, warning of the danger of country television stations merely becoming satellite-relay stations with no local content. Amalgamated Television Services Pty Ltd (then Channel Seven Sydney) opposed the plan and the professional lobby group for commercial television, FACTS, of which Publishing & Broadcasting Pty Ltd was a member, rejected the scheme. Under the weight of this pressure, PBL submitted a further proposal to the Task Force in which it maintained that its initial proposal was a firmly held long-term plan, but presented a revised case for a more modest extension of the number of commercial television stations.

Despite all the problems, the White Task Force recommended to Government the introduction of an Australian national communications-satellite system 'as early as practicable'. They argued for the creation of a new statutory authority, a satellite commission with the responsibility of 'assembling user requirements, specifying, planning, developing, marketing and operating the space segment.'[16] The satellite commission ought to be given 'considerable freedom' in marketing a full range of satellite-communications services, with the exception of 'domestic and international telephone, telex and telegram services directly to the public.'[17] These principal recommendations were not unanimous, with the representative from the Department of Finance dissenting from the central proposition to introduce a domestic satellite. The Telecom representative, though not dissenting from this, registered a protest against the creation of a new statutory corporation which would be independent of the monopolistic public common carrier, Telecom. The questions of cost and control of the satellite system were to remain the two most vexed issues during the ensuing years.

Harold White's Task Force essentially presented a 'blue sky' feasibility study. For them it was essentially a problem of how to prove use and benefit of a domestic satellite before it existed and operated. Their final recommendations were general and in no way constituted a detailed, convincing justification for the project. Their prime case was that a national communications system offered 'a range of improved, extended and new services over the broad frontiers of our national life.'[18] They wrote about 'the potential for applications and extensions of our national purposes, as yet unknown' and a 'new dimension to the communications capability within our country which cannot be totally comprehended by detailed considerations of its impact upon known services, service-by-service.' The notion of *services* is used in such a way that the implicit assumption is that change by this new technology is obviously beneficial to all. Australia, they argued, was ideally suited to this technology and they would never have proposed a domestic satellite if Australia 'with its 14 million people was a small country well latticed by a diversified terrestrial network capable of carrying all forms of electronic signals to all parts of Australia.'[19] They also pointed to the possible stimulus for the manufacturing sector, but warned that 'to be in a position to take advantage of some developments which may arise, perhaps in the 1990s or by the turn of the century, we need to acquire a national satellite capability during the 1980s.'[19] The subsequent tensions concerning the cost of the satellite were presaged in the initial work of the Task Force in 1978. Their all-in cost figure was an estimated $100 million to $165 million, subject to a plus or minus variation of 20 per cent.[20] J. P. Coleman of the Department of Finance, lodged a dissenting statement. He pointed out that Telecom had told the Task Force that unless there was a significant increase in the demand for television services it would be more economic to provide the desired increase in capacity through extension of the present terrestrial network. In his view it had not been clearly established that the necessary demand for television satellite-relay services would develop.

Following the 1977 national election, the new Minister for Post and Telecommunications, Tony Staley, had emerged as a firm supporter of the satellite. He tabled the White Task Force Report in Parliament on 27 September 1978, invited comment from the community, and set up a working group of senior officials from a number of Federal Government departments, chaired by members of the Department of Post and Telecommunications, to report to Government on 'all matters regarding a national communications satellite system.' Broadcasting interests were again unrepresented. Members of the working party, including senior officials from the Departments of Prime Minister and Cabinet and Finance, were aware that the Task Force's proposals were highly controversial, particularly in the areas of costs, how the proposed satellite authority would be constituted and possible usage by the broadcasting sector.

The key recommendation of the working party, that a 'multipurpose communications satellite capability should be established as early as practicable' and that an 'appropriate organisation be established immediately to develop the plans', was again not unanimous. The representative of the Department of the Prime Minister and Cabinet recommended deferment of any immediate decision, and the representative of the Department of Finance recommended that the introduction of a satellite should be put aside for two to three years when a further detailed study could be undertaken. Members of the working party all agreed on the 'potential benefits of a communications satellite system', but not on the 'how', or at what cost. A principal conclusion was that 'telecommunications in Australia would be enhanced by the *complementary* application of satellite and terrestrial systems and that a communications satellite system would overcome some of the most significant limitations in current telecommunications services.'[22] The areas which could benefit most were stated as provision of telecommunications services to remote areas otherwise unlikely to have such services, including telephony, the enhancement of television and radio broadcasting for Australia generally, and enhanced services for bulk high-speed data communications.

The Working Group admitted that it had 'difficulty in arriving with accuracy at actual costs of many aspects of a possible satellite system.'[23] They pointed out that all of their financial projections were subject to a significant degree of uncertainty, the most crucial elements of which were the extent to which a satellite system would generate additional revenue or merely shift revenue from Telecom, and how the tariffs to be charged would affect demand. They calculated costs in 1985 dollar terms and were aware that, since 80 per cent of potential users for the satellite system appeared to be government departments, these public 'on costs' must be taken into account. According to the Working Group, the overall public-sector net cash outlay for a system, including telephony and broadcasting services to remote areas, would be in the range of $200 million– $260 million over the period 1982-92 (in 1985 dollars).[24]

On the question of who would control the satellite, the Working Group reached similar conclusions to the Task Force. They opted for the establishment of a National Satellite Commission with overall responsibility which ought to be permitted considerable freedom in the provision and marketing of a full range of satellite-communications services 'with the exception of domestic and international telephone, telex and telegram services directly to the public.'[25] The representative from the Department of Finance dissented from this recommendation, with the suggestion that 'it might be better implemented by a largely autonomous subsidiary of Telecom than by a separate authority.'[26] Five years later the incoming Hawke Labor Government was to grapple with the same fundamental questions of the degree of autonomy that was desirable for the satellite authority.

In the broadcasting area, the Working Group steered a middle course between entrenched commercial interests as against opportunities for new broadcasting outlets. They pointed out that for technical and economic reasons the present broadband-transmission network was limited in capacity and quality, was expensive to extend, and would not in the foreseeable future provide services to many widely dispersed locations. With a communications satellite, they argued, high-quality television-distribution services could be provided to any location. However, in their view, the greater potential to use the satellite system in the production and distribution of television programs was with the ABC, rather than in the commercial sector, although they considered the commercial service could be substantially enhanced.

In their conclusions, the Working Party showed an awareness of the great potential for commercial-television distribution via satellite, but could not resolve the thorniest conflicts of interest between old and new, metropolitan and regional. It shelved decisions on the toughest questions relating to commercial television, but recommended that the Government 'ask the Minister for Post and Telecommunications to submit a proposal for future broadcasting policy in Australia taking account of the interests of viewers and listeners, the broadcasting industry and the capabilities of satellite communication.'[27] No Minister, in any Australian Government, has yet been up to such a task.

On 19 October 1979 Staley announced to Parliament that on the basis of the range of advice available to it, the Government had 'decided in principle that it is in the national interest to establish a national communications satellite system in Australia.'

His speech in the House and subsequent interview are wonderful examples of the rhetoric of a technological utopian. The system would 'bring major benefits to millions' . . . 'weld our scattered population into one nation'. . . indeed 'foster a sense of unity that will make us a great nation.'[28] The Government, the House was told, had undertaken 'a very full examination' of the issues involved and 'nothing could be gained by now delaying a decision in principle' to proceed. The Minister added that 'we have searched the world to glean the latest knowledge on space communications systems technology' and 'scrutinised overseas experience in adapting satellite technology to meet the burgeoning demand for communications services'.[29] Two crucial policy matters were ignored in the speech: namely, who would own and control the system, and firm cost estimates for the project – which would depend, according to the Minister, on resolution of the overall system design.

Major public-policy issues and pronouncements on technology have few forums for analysis and public debate. Press response was generally favourable with the *Australian* editorial (19/10/79) stating that 'in a land as vast as Australia it makes plain common sense' and the *Age* editorial (20/10/79) seeing the project as 'an imaginative one which will bring the benefits of space age technology to a great many Australians'. No such technocratic lines appeared in the *Financial Review* editorial (19/10/79) headed 'Pie in the Sky' which said bluntly that 'the Federal Government's satellite proposal is as unsatisfactory as the discussion which has surrounded it and the decision making process from whence it has come.'

In November 1981 the new Minister for Communications, Ian Sinclair (Staley had retired from Parliament), announced that the system would be officially known as Aussat. It was envisaged that the company would operate on a fully commercial basis, provide services at non-discriminatory tariffs, pay taxes and generate dividend return to its shareholders. The authorised share capital of the company was $100 million, divided into 100 million shares of $1 each. Under the Fraser Government the sole beneficial shareholder of the company was the Commonwealth of Australia, but Aussat's Memorandum of Association directed the company to convert to a public company, as soon as practicable, with up to 49 per cent of the issued share capital 'owned by persons other than the Commonwealth of Australia or its authorities', to be listed on Australian stock exchanges.

The intention to open up satellite ownership to the private sector was from a Government which embraced the deregulatory mood in broadcasting and telecommunications, and whose intention was to minimise public-sector spending. A Government with enmity towards Telecom was seeking change in the structure of Australian telecommunications. The interim management through OTC, to develop and design the technical specifications for the satellite, changed to a thirteen-member board nominated by government, with representatives largely from private enterprise, including the major networks and commercial regional television interests. Predictably, Telecom was left off the board of Aussat; and the ABC, to be designated a major satellite user, was also unrepresented. However, senior staff members from five major Australian commercial organisations, including the Packer, Fairfax, News and Herald groups, became directors of Aussat Pty Ltd.[30] Under the Fraser administration, Aussat Pty Ltd had only two public-sector representatives, the Deputy Secretary of the Department of Communications and a Commissioner of OTC, on its all-male 13-member board. The foundation board, of course, would have given way to a partly-elected board had the public company ever been floated under the Fraser Government.

The Hawke Labor Government inherited a legacy of unresolved policy questions surrounding the satellite project when it came to office on 5 March 1983. The new Minister of Communications, Michael Duffy, the fifth Communications minister in six years, gained the portfolio.

He had not 'shadowed' this area of ministerial responsibility in Opposition and soon found that he had to confront complex, thorny satellite issues. In the wake of the Davidson Report and Labor's election-campaign promise to support Telecom's strategic position as national common carrier, the Telecom union leadership of the ATEA mounted a $20,000 publicity campaign to scrap the national satellite project. At a time when many groups in the telecommunications and broadcasting industries were frustrated that the new Minister chose to make himself inaccessible to them, the ATEA union leaders were able to plant grave doubts about the viability of the project with Duffy and other senior ministers. The new Minister, in May 1983, requested that the Department of Finance review the costings of Aussat and he argued that the project should be self-financing, that it must not become a drain on Commonwealth funds, and that Aussat should not engage in 'destructive and damaging competition with Telecom's terrestrial network services.'[31] However, whilst doubts about viability were firmly implanted, no member of the new Government publicly opposed the continuation of the project.

In mid-August 1983, with satellite issues now on the national political agenda, the Labor Government Cabinet and Caucus formally endorsed the national satellite project, though no details about the satellite authority or operational arrangements accompanied their in-principle decision to proceed. Clearly, fear of the political backlash of killing the project, which would have led to the accusation of wasted public funds for the pay-off, estimated to be at least $150 million, outweighed detailed consideration of the long-term financial implications. Cabinet, in fact, in making their decision, did not receive detailed costings from the Minister of Communications.[32] Duffy had become increasingly aware of some strong support for the project from areas he could not afford to ignore, particularly the Western Australian and Northern Territory Governments.

In his early days as Minister for Communications, Duffy had in mind a trade-off for Telecom management and unions – the satellite project would continue, but Telecom could run it. Speaking at a Victorian ALP State Conference in June 1983 he expressed a personal view that Telecom should manage the satellite project to minimise the risk that Aussat could become a competing common carrier.[33] Telecom management quickly responded by putting several ownership alternatives to the Government – total integration of the satellite project by them, or Aussat as a wholly owned subsidiary of Telecom, or part-ownership of the system either with Government or private enterprise.[34] At this stage the Hawke Government stayed with the notion of an independent satellite authority with 51 per cent public ownership and 49 per cent private shareholding. New Treasurer, Paul Keating, endorsed this satellite-ownership principle of the Fraser administration in both his May 'mini' 1983 Budget and the August major annual Budget address. Duffy did, however, move to change the composition of the Aussat board of directors. He dropped all members who represented commercial television and newspaper organisations on the grounds of potential conflict of interest, since their companies were expected to be large customers of Aussat. As an important symbolic gesture of Labor's compatibility with Telecom, Bill Pollock, Managing Director of Telecom, became a director of Aussat.[35]

On 15 November 1983 the Labor Government made its most detailed Parliamentary statement about ownership and operational aspects of the domestic satellite system. In a policy switch, Duffy announced that Aussat was now to be retained as a Commonwealth-owned company and Telcom was to be invited to purchase up to 25 per cent of Commonwealth shareholding, thereby revoking the earlier decision to offer 49 per cent of the Commonwealth asset to private shareholders. The justification was that this policy would provide that 'Aussat and Telecom co-operate to bring about the co-ordination of satellite services with terrestrial services.'[36] This way the government maintained full public ownership of the domestic satellite system, but with Telecom's financial involvement, Duffy was able to overcome Treasurer Keating's concern about any further additional burden on the national deficit. Obviously a 49 per cent private float would have reduced the level of government financial commitment, but this was rejected in favour of a 75 per cent government – 25 per cent Telecom financial partnership of Aussat.

This was essentially a compromise decision. On the one hand, full Telecom ownership of the satellite system would have been met with hostility from the business community, fear from Aussat, and opposition from some sections of the Labor Party. On the other hand, a model which allowed considerable private-sector involvement, and a free hand for Aussat, was clearly unacceptable to the union base of the Labor Government. It was a tight decision in the Labor Party for Michael Duffy. On 15 November 1983 Caucus voted to deny Telecom control of the satellite, by a narrow margin of 43-40 in favour of Cabinet's decision to keep the satellite system in the hands of Aussat, with a 25 per cent Telecom shareholding.[37] Duffy had, of course, declared earlier his personal support for integration with Telecom, and so he was placed in the position of having to oppose his own arguments to Caucus in favour of Cabinet's decision.

On operational matters, Duffy's Ministerial statement argued that the Government should only establish broad guidelines for access by broadcasters to the satellites, but not be involved in the direct allocation of transponders which ought to remain a matter of commercial negotiation between Aussat and prospective users. On the question of ownership of earth stations the Government decided that private-sector users of the satellites may 'own and operate earth stations to access the satellite and own terrestrial links to connect their earth stations to their premises', but they would 'not be permitted to sell spare transponder capacity, nor allow third party traffic'.[38] This was a significant decision because it gave approval to those private companies who could afford to buy their own earth stations to by-pass Telecom for some private telecommunications traffic, including long-distance phone calls and data transfer. Duffy had again shifted his position from doubts he had expressed earlier on a *Four Corners* program about 'inadequate control' if private-sector users were allowed to transmit as well as receive satellite signals through their own earth stations.[39]

Where the Fraser administration had permitted the nucleus of new alternative private telecommunications networks by granting private earth-station ownership via Intelsat satellites for television channels TCN-9 and ATN-7, the Hawke Government had now opened the way for diversion of traffic from Telecom by allowing private earth-station ownership to access Australia's domestic-satellite system. The beginnings of the apparatus for an alternative national telecommunications network – the Fraser strategy – had been endorsed by the Hawke Government.

9 Aussat: The Politics of Transponders

The political decision-making process which led to the introduction of Australia's domestic satellite system was essentially conducted behind closed doors, and between competing vested interests, with the ultimate decisions primarily made by technocrats. The group that deserves the most commendation for attempting to generate wide public debate was the Australian Telecommunications Employees Association (ATEA): not governments, or a government Department of Communications, or the satellite authority, but a trade union. There are few forums in Australia where the complex issues concerning the introduction of new technologies can be debated at length, and virtually no independent research authorities have been commissioned to investigate major questions concerning the desirability, advantages or impact of this technology for the people of Australia. The crucial questions now are how the satellites will be used, and who will have access to this telecommunications alternative. Its opportunities must be maximised for public benefit.

Design of the Australian Satellite System

Given that the satellite project was expected to pay its own way, the design exercise was to accommodate as many different kinds of potential customers as possible, all within a single national satellite communications system. As a consequence, a flexible hybrid system emerged from the designers, with transponders of different power. In the planning stage of the project, the major uses were identified as direct television broadcasts to homesteads and remote communities, television relays between major cities, digital data transmission, voice applications for remote areas, centralised air-traffic-control services and maritime radio coverage.

Australia's initial system will consist of two solar-powered satellites in geostationary orbit with a spare on the ground. The location of the two geostationary satellites, 36,000 km above the equator, will be at 156°E and 164°E longitude, with 160°E reserved for the third. For the launch of satellites long advance bookings are required, and Australia made reservations with three different launch authorities. One was with the Delta expendable launch vehicle of America's NASA, another with the European organisation Ariane, but the final choice was for the American re-usable space shuttle. Australia's first two satellites are scheduled to be launched from Cape Canaveral on board NASA's Space Shuttle in August and October 1985. The satellites are the spinning-drum model: 6.6 m tall, 2.2 m in diameter and weighing about 1250 kg at launch. They do not function forever in space, but have a life expectancy of seven or eight years. European company Arianspace has been contracted to launch Aussat's third domestic satellite, with launch options from mid 1986 through to mid 1988.

Each satellite has 15 transponders, four of which will have a power output of 30 watts each, and the other 11 transponders are 12 watts each. Three satellites, with transponder capacity each of 4 x 30 watts and 11 x 12 watts, is substantial capacity to serve a nation of 15 million people. The frequency band for the uplink will be 14-14.5 GHZ (ground to satellite) and 12.25-12.75 GHZ bank for the downlink (satellite to ground). The satellites will be monitored by two control stations, located in Sydney and Perth, known as Tracking, Telemetry Command and Monitoring (TTC&M) stations.[1] The three satellites and both control stations are to be supplied by Hughes Communications International, the world's major manufacturer of communication satellites.

The Aussat transponders are focused in either a national beam, which covers all of the Australian continent, or in zonal or spot beams. The four spot beams each cover Queensland or South East Australia or South Australia and the Northern Territory or Western Australia. Remembering the general rule that the higher the power output from space the smaller the receiving antenna (or dish) on the ground, the system was designed to enable remote communities to receive some television programs, preferably to pick up on a receiving antenna that would cost less than $1000. To meet this objective those transponders with the greatest output – 30 watts – are also concentrated into a limited area of the continent.

These 'spot' beams for the Australian states, which collectively cover the whole mainland, were designed to offer radio and television services to isolated communities as a Homestead and Community Broadcasting Satellite Service (HACBSS). In an early announcement about guidelines for satellite users, and consistent with the notion of equity of communication services, the then Minister of Communications, Ian Sinclair, referred to the 'transmission of ABC television and a minimum of two radio services by relay to regional transmitters, and by direct transmission from Aussat satellites to remote homesteads and communities embracing about 300,000 Australians.'[2] The use of the second HACBSS facility, the highly sought spot beams of the second satellite, remained an unresolved problem.

The eleven lower powered transponders on each satellite were designed to feed a national beam to cover the Australian mainland, Tasmania and some offshore territories. From the initial ministerial statement on satellite services, and the early statements from Aussat Pty Ltd, it is possible to detail a range of telecommunications for which these transponders could be used:

★ Interstate production, distribution and exchange of commercial and national television and radio programs.
★ Low-power direct broadcasting of television and radio programs to isolated areas.
★ Relay of television and radio programs from central studios to regional stations for re-broadcast.
★ An automatic telephony service to remote Telecom subscribers and the provision for back-up emergency and temporary main-trunk-route telephony services to assist in reducing congestion in areas of presently overloaded mainline-trunk-telephone demand, or to restore services in an emergency.
★ Aeronautical, maritime and land communications to meet Department of Transport operational and safety requirements.
★ Voice, data and other telecommunications services to meet requirements of commercial, scientific and mining interests and various State and Commonwealth Government departments and authorities.
★ Telecommunications services to Norfolk Island, Lord Howe

Island, and the oil and gas rigs such as those in Bass Strait and on the north-west shelf of Western Australia.

★ Education broadcasting of television and radio programs via satellite, broadcast teletext services, specialised learning materials, and an improved School of the Air.

★ Quick establishment of communications facilities using transportable earth stations for emergency and disaster-relief situations.

★ High-speed computer communications and digital data services for public and private-sector organisations such as banks, data information and computer-service bureaux and Government departments.

★ Provision of domestic telecommunications and television and radio broadcasting services for Papua New Guinea.

This is nothing more than a check list of possible uses, but it indicates a broad spectrum of prospective customers for Aussat.

The issues surrounding the forms of ownership and functions of earth stations were gradually resolved during 1982 and 1983. It was decided that Major City Earth Stations (MCES) would be located in eight places – each of the state capitals and Darwin and Canberra – all owned by Aussat Pty Ltd. These were to be both transmitting and receiving earth stations for radio, television, data and telephony signals, both to and from the satellites. On the landscape they appear as single-storey brick buildings with one or two dish antennas and a radio-relay tower with associated facilities. Mitsubishi Australia Ltd won the contract for supply of the eight earth stations, with half the contract value of about $16 million to be spent in Australia. These are 'gateway' access points to the satellites via customer-interconnect links for those users who do not have their own earth stations.

In addition to the earth stations run by Aussat, a number of private users can choose to provide their own communication facilities, and lease space-segment capacity through Aussat Pty Ltd. The Hawke Labor Government approved private-sector usage of the satellites to own and operate earth stations, and to own the terrestrial links to connect their earth stations to their premises.[3] Several categories of earth stations with antennas ranging in size from 1.2 metres to 18 metres, with obvious correlation for cost, will be used. Australia will, of course, be a booming market for satellite earth stations, but as we have seen earlier, the opportunities for indigenous Australian companies to reap the rewards of this new sunrise industry will depend upon the successful application of policies designed to build a national technological infra-structure.

It is worth noting that the popular Hughes-built HS 376 satellite for Aussat is becoming an obsolete model. The replacement model, expected to be operational by 1988, will offer at least 50 per cent increased capacity for comparable weight.[4] The new Hughes version will offer a significantly improved direct-broadcasting capacity. This raises the perennial question of how long to wait to get better and more cost-efficient new technology.

Selling Satellite Capacity

An inherent characteristic of a communications satellite is that it has a fixed capacity, or a specified number of transponders, from the beginning to the end of its life in space. Unlike the terrestrial network, where it is comparatively easy to gradually add components to the system, the total capacity of a satellite remains constant during its lifetime, provided there are no technical failures. This presents a challenging marketing problem in that from virtually the first day of its operation a satellite possesses its full capacity. Aussat Pty Ltd was conceived as a commercial operation, whose task was to sell transponder capacity to users in a way that would guarantee the financial viability of the project. The usual method of selling access to the communication facilities of a satellite is on the basis of an annual rental fee per transponder. Aussat's marketing task is to find the best customers in Australia to give major long-term commitments to use the three domestic satellites.

Selling the satellite capacity of Aussat is not, however, solely a commercial marketing operation. The setting of market parameters is inherently a political decision, to be made by the government of the day. What, for instance, is the relative importance of the forms of transmission – broadcasting, telephony or data? Will, for example, the satellite authority be permitted to compete openly in the transmission of telephony? Should, for instance, the prized high-powered transponders be allocated for broadcasting, and if so, to commercial or national or community broadcasters? In its early days Aussat was handicapped by its dependence on important government communication-policy decisions, which took so long to emerge. It was remarkable that the Government which commissioned the satellite project, at such substantial public expense, the private-enterprise Liberal-National Party Fraser administration, could so consistently shelve the decisions which were necessary for Aussat Pty Ltd to operate a successful commercial venture. It was rather like floating a new company, pouring money into it, but having little idea of which products to market, or how. Access to the satellites was a complex and intensely political debate which the Fraser administration left unresolved by the end of its time in office.

Primary and Secondary Users

It is useful to categorise as primary users those organisations who indicated a firm commitment which was within government-policy guidelines, and as secondary users those who had expressed interest, but where the issues of how they might be involved, and at what cost, needed resolution.

Primary users, or services, were:

1. Broadcasters: The Australian Broadcasting Corporation for television and radio, and the commercial/television networks.
2. The Department of Aviation for air-navigation services.
3. Telecom for remote telephony services and route diversity.
4. The ABC's remote-area broadcasting service.
5. Private-sector data and information services, including private networks.

Secondary users, or services, were:

1. Community broadcasters or radio programs distributed through the Public Broadcasting Association of Australia.
2. Educational programs, by state-based educational organisations or outback educational services such as the School of the Air.
3. The Papua New Guinea Government, for television and telecommunications.
4. New broadcasting systems, such as radiated subscription television services, or pay television.

While the Packer group had been the catalyst for the introduction of a domestic satellite, the conflicts within the industry prevented commercial broadcasters from gaining guaranteed access to prime transponder allocation. There was nothing to prevent normal commercial negotiation between Aussat and a commercial broadcaster for access to the low-powered transponders, but Government policy did not support an entrenched position for the networks. The resolution of conflicts between metropolitan and regional broadcasters was too difficult for a coalition of Liberal-National Country Party interests.

On 6 May, 1982, Ian Sinclair, Minister for Communications, at last presented Parliament with a detailed statement on the domestic satellite project. Only three groups were assured of access – the Australian Broadcasting Commission, Telecom and the Department of Transport. The statement did make references to commercial and business users, including commercial broadcasters, but did not specify how these groups would operate with Aussat. It was remarkable that first prize on the satellite system – guaranteed access to the set of high-powered transponders on the first satellite – should be given to the Australian Broadcasting Commission by a Fraser administration so antipathetic to that organisation.

Satellite technology is not new to the ABC, which had used circuits of the Intelsat system in conjunction with Telecom's terrestrial network to provide an outback broadcast service which began in 1980 called the Remote Area Television Scheme (RATS). The receiving earth stations for RATS need to be larger than those to be used in the domestic system, with an antenna diameter of 6.5-7.5 metres, at an average cost of $125,000, with a transmission radius of 10-15km.[5] Translators re-broadcast programs for reception by homes well beyond the range of the existing terrestrial broadcast system. In addition to these official stations there are some private earth stations usually operated by mining companies, at an average cost of about $7000.[6] These stations receive television transmission, but have no re-broadcasting capability. RATS extends from Thursday Island, Weipa and Gove in the North East, to Meekatharra, Exmouth and Paraburdoo in the West and reaches almost 70,000 people.

Outback broadcasting via Aussat is to become a new task for the ABC. The high-powered (30 watt) transponders on the first satellite, used in conjunction with spot-beam antennas, were designated to the ABC to provide a Homestead and Community Broadcasting Satellite Service (HACBSS). This will provide a direct broadcasting service to individual homesteads and isolated communities, which may also be received and re-transmitted by regional ABC stations over terrestrial transmitters. The signal level of the HACBSS transponders, operating through a spot beam, will require a receive aerial of between 1.2 to 1.5 metres in diameter, at an estimated cost of about $1000. The intention was that by allocating the concentrated spot beams on the first satellite in this way, the cost of receiving antenna could be brought within the purchasing capacity of isolated homesteads and small clustered rural communities in which several homes could be linked to a single receiving dish. After these once-only costs have been met by the inlanders, this brings ABC broadcast services to the outback, free of direct charge.

The Dix inquiry into the ABC argued that HACBSS would provide a much improved amenity to people living in the remote regions, but would have little impact on people in the more populated parts of rural Australia 'unless Government agreed to fund the additional capacity necessary to deliver two additional radio networks to these areas.'[7] In its first budget, the Hawke Government accepted this recommendation from Dix and the ABC management. The Minister for Communications, Michael Duffy, announced that the ABC could develop a second regional radio network with more than 40 regional studios, of which a quarter would be new, at an estimated cost of $40 million over the next five or six years. Growth for Aunty!

As well as new outback services, the key satellite exercise for the ABC has been to examine the comparative costs of transmission via Aussat with costs by the traditional Telecom terrestrial network. It is highly likely that ABC television will become more national in its transmission methods after Aussat is operational. The dispersed studios of the ABC currently originate program material and exchange it by means of terrestrial sound and vision circuits, and by tape exchange by aircraft. It is likely that by the end of 1985 Aussat will be the major transmission facility for ABC television, with the majority of programs emanating from one point, the Sydney Gore Hill studios. This will involve the redeployment of some technical staff in ABC offices throughout Australia. The closures of regional television centres in Rockhampton and Townsville seem inevitable.

Other possible uses of Aussat by the ABC are dependent on funding. In April 1984, ABC management expressed its interest in using a total of 9.5 transponders on the satellite system, composed of four high-powered transponders for its HACBSS, one low-powered transponder for South Australia (to enable Northern Territory and South Australian programs to be separate), three low-powered transponders for closed-circuit television interchange and 1.5 transponders for other radio and data-transmission services.

There are two major components to the new communications technologies for any organisation – access to new delivery systems and opportunities for program development. The ABC will have guaranteed access to the satellite, a national distribution system on the readily available VHF spectrum, without direct charge to Australian viewers. This is merely comment about assured delivery – it is another issue as to how the program services will be funded and developed.

Public broadcasters throughout the world have been concerned that the development of new communications technologies would undermine their pre-eminence and greater priority would be given to new private-sector technological developments. The trend in domestic satellite developments throughout the world, and particularly in the USA, is for large private corporations to be responsible for the costs of the launch, services offered and business risk involved with this new technology. If ABC management satellite plans are all funded, it would be the largest dedicated access ever assured to a single public-sector broadcasting organisation in the history of communication satellites throughout the world. At the very least, the ABC is assured of HACBSS 1, and a prime place in Australia's national domestic satellite system.

One of the staunchest supporters of the national satellite project, the Department of Aviation, is expected to be the second-largest user, after broadcasters, of domestic-satellite capacity. The satellites will offer improvements in the quality and coverage of communications between aircraft and airtraffic control and flight-service units, especially in remote areas. At present, much of Australia's aeronautical radio traffic on overland outback communications, for domestic or international flights, uses HF radio signals. These are shortwave frequencies which cannot provide the direct pilot-ground operator communication which is enjoyed on the major routes using VHF radio signals. HF radio is crackly, affected by solar storms, and prone to fading without warning. A daily Qantas flight between Sydney and Singapore, with its long inland-Australia route, uses a 'bits and pieces' system of both VHF and HF radio, requiring changes of radio equipment during the flight.

The Aussat satellite system will provide high-quality direct pilot-to-ground-controller VHF communications over much of the Australian continent, and airtraffic centres will be interconnected for both voice and data exchange. The new VHF network would employ duplicate facilities to operate independently through each of the two domestic satellites. Airtraffic centres will be able to choose the better of two independent signals, which ought to ensure greater safety in the event of atmospheric interference, or if one satellite or its ground equipment failed. The Department of Aviation announced its commitment to the system in August 1982 and will use two low-powered transponders on each of the two Aussat satellites. The Department of Aviation network will have a total of 99 ground stations, with 53 smaller ground stations in remote locations and 45 larger earth stations at existing sites.[8]

Telecom will be involved in implementing the key political justification for the introduction of satellite technology, the promise of greater equity in the distribution of communication services throughout Australia. There is no doubt that the provision of a telephone service to the people of the outback is an extremely appealing prospect to those without telephones. In 1980, a report commissioned by Telecom found that 'pastoralists and others generally would be prepared to pay significantly more for the telephone than for other services such as broadcast television or radio mobile communications', whilst for Aboriginals there 'is a clear need for contact with non-Aboriginals, to ensure maintenance of personal health, legal rights, food supplies, contact with kin and entertainment.'[9] Yet two things must be made clear at the outset about the provision of a telephone service to isolated Australian communities – it is an extremely costly service to provide, and a domestic satellite is not the only technology by which telephones can finally reach inland Australia. Unfortunately some people in the outback mistakenly believe that the satellite itself will somehow instantly provide a telephone service, at minimal costs to them, soon after the system becomes operational.

Telecom have, in fact, opted for additional outback telephone services mainly through the extension of a terrestrial network system based on a technology known as the Digital Radio Concentrator System (DRCS). With this technology the speech signal from the telephone is converted into a digital code, and a number of telephone lines can be connected to a telephone exchange, using a single system radio carrier. The system uses UHF radio links from the nearest existing telephone line, and by using repeaters at about 50 km intervals the subscriber range can be increased to 450 km. It will allow the full range of network services to be provided, such as STD and ISD telephone services, telex, low-speed data and leased lines to customers who live up to several hundred kilometres from the telephone exchange.

While a 1980 estimate suggested a cost per subscriber to the DRCS of $8700[10], the Minister for Communications estimated, late in 1983, that 'the installation cost to Telecom for the majority of DRCS services was expected to be between $15,000 and $60,000 per customer, with an average cost of about $20,000'.[11] Telecom has recently been involved with DRCS proving trials and the first production systems are expected to become operational in 1985.

It is anticipated that about 3000 outback subscribers can be linked by this method to the telephone network, but an additional 200 in the extremely isolated areas could only eventually be given services through the use of the domestic satellites. If the DRCS costs to outback consumers appear to be high, they should be compared with the costs for Telecom's satellite telephone service, as estimated by the Minister for Communications in October 1983 – the installed cost of a remote earth station and associated terminal equipment of $60,000, as well as the annual charge for the rental of the transponder, to which must be added costs of the main earth station and special exchange equipment.[12] Again, not widely understood in the Territory was that small earth stations, to be used for the HACBSS for radio and television, would not be able to be used for a satellite telephone service as well. This would require an additional earth station, which would have to be much larger than a HACBSS 1 dish.

Speaking at a conference in Alice Springs, organised by the Northern Territory Government, in July 1983, to ensure the retention of the satellite project under the new Labor Government, Roger Banks, Director of Business Development for Telecom, declared that Telecom's major long-term planning was to see all Australian customers connected to an automatic national network by 1990. If this assurance is met to include *all* Australians, it will be one of the most expensive inland telephone systems in the world. However, in addition to this major goal, the satellite system would be useful, according to Telecom plans, to provide emergency restoration of services to areas isolated from the network in a time of disaster, and the opportunity to provide service at short notice, or for a limited period of time, to mining or exploration sites.

Telecom has also developed a domestic-communications satellite service called the Iterra Network Service, which is expected to be operational from January 1986. Iterra is Aboriginal for 'be quick', and the service will provide telephony, data and text facilities connected to the national switched telephone network via satellite for organisations operating in the more remote areas of Australia. In addition to services for the outback, it is probable that Telecom will embark upon another major usage of the satellite, the provision of itinerant and back-up capacity for Telecom's main trunk-route services.

The last of the group of primary users are private-sector companies who will use Aussat capacity for voice and data transmissions through their own private networks. A satellite lends itself admirably to use by organisations who need to distribute information to branch offices or to clients widely dispersed throughout Australia. The decision was taken by the Labor Government that companies could own and operate their own earth stations for their own needs, and the uplinking of the satellite was permissible.

That was a popular decision with the private sector but contentious elsewhere, in that Telecom and its unions saw it as an acceptance of some privatisation of telecommunications, and the beginning of cream skimming of Telecom revenues. Data and information services can be distributed through low-powered transponders which carry associated high-cost earth stations. Few groups have given details of their plans in this area, but the number of corporate users who have sufficient demand for their own expensive earth stations is expected to be small. The list might include leading computer companies such as IBM, financial organisations such as major banks and large insurance companies, and media organisations who specialise in information services, such as Australian Associated Press (AAP).

AAP is probably the largest private-sector operator of a private telecommunications network in Australia. It is a major supplier of news sources to newspaper, radio and television companies throughout Australia, and it is jointly owned by the major media groups. AAP has a stake in another major news supplier, Reuters, with whom it operates jointly in some areas. AAP says that it provides information to 'the professional user who requires 24 hours a day permanent connection to one or more data bases with specialist services and minimal retrieval times . . . for a one-up fee for the equipment and continuous system access.'[13] Their specialist information services include a domestic and international money-market service, an Australian and overseas stock-exchange prices retrieval system, and a racing-data base called Form Guide (or Neddy bank) which covers 108,000 horses and 172,000 racing events in Australia and New Zealand during the past seven years.

AAP intends to extend its range of information services under Aussat and lobbied for private-sector ownership of earth stations. It warned the Government of doubts about the financial viability of the satellite project and argued that if private uplinking was prohibited, then AAP would reduce its formerly guaranteed commitment to Aussat, to take one full low-powered transponder, to only 23 per cent of a transponder.[14] AAP planned a $30 million expansion program based on satellite technology, which it claims will generate 200 jobs. It intends to install multi-directional microwave transmitters in each capital city in Australia and up to 58 receive-only 2.5 metre dishes at country provincial centres, for information distribution via Aussat at the beginning of 1986. For AAP, the domestic satellite system will encourage the extension of its services geographically, and the addition of new information-services aimed at markets such as tertiary institutions and schools, libraries, hospitals, doctors, chemists and small businesses. AAP has devised an ambitious strategy to become a key information-provider in 'the satellite age'.

The Second HACBSS

The politics of access by broadcasters to the second set of high-powered transponders, or HACBSS 2, was one of the most vexed and delayed policy decisions ever made in Australian communications history. In the planning phase of the Aussat system it was envisaged that the second satellite would only be used as a back-up in case of the failure of transponders on the first satellite. However, by 1980, the impressive reliability of satellites in space after successful launch and correct orbital placement was widely acknowledged. There were several pressures which favoured the leasing of the second set of high-powered transponders. Obviously, the prospect of additional revenue was appealing to Aussat Pty Ltd, and several broadcasting organisations expressed interest in using this spare capacity on a pre-emptible basis. A broad range of options was laid out by R. Lansdowne, Secretary, Department of Communications, in a speech early in 1980 when he explained the various ways in which a second Homesteads and Community Broadcasting Satellite Service could be used:

★ a commercial service licensed to an *existing* licensee on a national basis, or licensed to four separate existing licensees on a zone-by-zone basis
★ a commercial service licensed to a *new* licensee on a national basis, or four new licensees on a zone-by-zone basis, providing alternative programming to the existing terrestrial commercial services
★ a nationwide subscription-television service, licensed to private enterprise, or operated by the ABC
★ a second ABC television network
★ a national multi-cultural television network
★ an educational television network, perhaps also incorporating other forms of special-purpose television
★ various combinations of the above.

These matters were to be resolved by the Government, Lansdowne explained in his speech, later that year, 1980.[15] They were in fact never resolved under the Fraser administration and were handed on in 1983 to the Hawke Government.

The Department of Communication in 1981 argued for an evolutionary development of policy for direct broadcasting services, and established a Task Force on Broadcasting Development within the department to look at how a variety of hybrid services might be accommodated on HACBSS 2.[16] They proposed a flexible approach to the use of high-powered transponders where each of the four zones could be served by two television services and at least four radio services, embodying national, commercial, multicultural and 'special purpose' components. They proposed that a system could be devised where particular services could be offered at specified times, such as multi-cultural television on one evening, and special-purpose television on the down-time during the day. Whilst there would have been many thorny choices to be made between vested interests in particular time allocations, the admirable intention was to have as many different groups as possible accessing the satellite. For HACBSS 2, this proposal departed from the notion that it had to be a fixed 'point to point' satellite service, run by one operator, to the notion of sharing a transponder on a time and frequency basis. Hence, there would be a wider range of users and a more diverse range of services.

The project also had the great merit of becoming an experimental direct broadcast satellite (DBS) service for Australia. However, the intensely self-interested commercial-television industry refused to co-operate in this imaginative collaborative satellite venture. It thereby lapsed. As the Australian Broadcasting Tribunal later observed, 'the established interests seemed to have wished to reserve the field of DBS to themselves until such time as it became profitable for them to enter it; and meanwhile sought to close it off to new entrants who might have been prepared to undertake the initial risk investment.'[17] Subsequently, the metropolitan commercial networks pushed for their own exclusive access to HACBSS 2 for their own national programming distribution. A design change in 1982 made it technically possible for three of the high-powered transponders on each satellite to be switched to national beams. This had obvious commercial appeal to both the government and to Aussat, who were committed to decisions which would guarantee the eventual profitability of the national domestic satellite system. A problem with the imaginative TFBO alternative was that small users have small pockets.

There were always great tensions within the commercial-television industry regarding possible changes that domestic-satellite technology would bring to the industry. The general concern has been that satellite technology would promote further influence in extensions of commercial-network involvement into new outlets, and further the problems of unfortunate high levels of concentration of television ownership and control in Australia. Regional-television operators felt especially threatened by what they saw as a trend towards a national commercial-television system in the satellite age, where regional commercial-television channels could eventually be replaced by slave relay stations, run by the big city networks.

The concept of networking needs some explanation before the

regional: city arguments can be canvassed. Networking is a co-operative business arrangement between a group of television channels for the acquisition of local and overseas programs, for program production and distribution and for promotional purposes. Under Section 92 of the Broadcasting and Television Act it is only possible for one company to control a maximum of two television channels (after that their level of prescribed interest must be restricted to less than 5 per cent).

Hence the Seven, Nine and Ten networks are amalgamations of differently owned business interests. In a network, each member station retains nominal responsibility for its own programming but certain programs, such as major sporting telecasts, serials and quiz shows, are produced or purchased on behalf of the network and distributed to member stations, either on film or videotape, or direct by landline. Each station pays a charge for individual programs based on its share of the total network audience. Although centralist in its structure, this networking arrangement allows, in theory, local programming autonomy which means that stations can attract local advertising.

One of the areas of business insecurity relating to the satellite was that the present model of networking might be replaced by a different type of networking based on the American model. In this scheme, the big US networks carry a fixed schedule of programs which is relayed in its entirety to all network stations. Advertising time is sold by the network, and affiliated stations do not pay for programs but receive a share of total advertising revenue after all network and program costs have been deducted. Hence, affiliate stations have little incentive to carry local programs, since they must cover the full costs of such programs by raising their own advertising to cover the costs. Consequently most affiliate stations in the USA, though independently owned and licensed, are predominantly repeater stations for the network.

The fear by Australian regional-television management was that the effect of satellite networking would make the American commercial model much more commercially attractive to the networks than the present co-operative Australian model. Speaking at a Canberra conference in 1979, Ken Stone, Secretary of the Independent Station Satellite Committee, outlined fears for the demise of independent regional television with a communications satellite:

What worries us is that the sorts of recommendations which are being bandied around, and are getting a lot of sympathy, could lead to the demise of independent television stations in the regions, the end of local television expression in the regions in the arts, the news and public affairs and community service for regional citizens, the monopoly control, if not the monopoly ownership of regional television stations by media interests in the dominant capitals of Sydney and Melbourne, a considerable amount of unemployment, probably about 800 to 1,000 jobs of skilled television people and the elimination of those jobs are prospects for the fledglings who will emerge eventually from the Film and Television School. And of course it could represent a serious reduction in regional identity, regional motivation and enterprise, and regional will, which after all is a substantial part of the spirit of federalism and the philosophical base of decentralisation.[18]

Ministerial guidelines for HACBSS 2 were announced early in the term of the Hawke Labor Government by the Minister for Communications in November 1983.[19] It was announced then that HACBSS – Homestead and Community Broadcasting Satellite Service – would only be offered by the ABC. It had been decided earlier by the Fraser Government that the first set of 30 watt transponders should be allocated to the ABC for direct broadcasting to the outback, but Labor then made this an exclusive ABC satellite service. The Minister argued that the advantages of satellite technology were in making programs available nationally, and that satellites were best used for program assembly and distribution between broadcasters.

The November 1983 policy was that 'commercial broadcasting uses of all transponders will be confined to national beams and will be encoded.' The networks would be able to offer syndicated distribution, or 'satellite program services' (SPS) available for sale to other commercial or regional broadcasters. Resolution of problems inherent in this decision was left to the Australian Broadcasting Tribunal, which was given the task of monitoring the use of SPS through formal authorisation of the commercial television licencees involved. The Minister instructed the Australian Broadcasting Tribunal to conduct an inquiry into 'matters such as program standards, hours of service, control of stations, local content, exclusive contracts and the proportion of national to local advertising', to report by May 1984.[20] This 'SPS inquiry' was essentially about networking and localism. The concept of localism in television embraces local control of the channel, with locally originated programs and local advertising. It also involves regional decision making about the acceptability of overseas and Australian productions to which a local television station can purchase broadcast rights. Regional-television operators have traditionally benefited from monopoly in that they have been able to buy program rights for their own channel from the range of programs offered by the three commercial networks. They can go 'cherry picking' for their one channel from the programs offered by three possible suppliers, a position which is to their advantage as a buyer's market. What satellite technology offers is that regional commercial-television markets could be opened up to receive the programs from the three networks in real time, assuming that the number of regional channels was increased to accommodate the additional satellite signals.

The method of buying satellite capacity is for a network to pay for the exclusive use of a transponder. The cost of this leased capacity remains the same annually, regardless of the amount of programming relayed through the transponder and the number of stations to which it is relayed. Hence, the networks have a strong economic incentive to maximise transponder usage for the delivery of their programs. Moreover, regional-television channels rely on advertising revenues which have been generally on a 70-30 basis of national to local advertising clients.[21] Networks, then, have a further economic incentive to integrate national advertising with the satellite program feeds. Centralised networks would command a greater pull over advertisers. Hence, extensive use of the domestic satellite system by networks must lead to greater commonality in the scheduling of programs, at the expense of local programming autonomy, and increased use of integrated national advertising, which must undermine the commercial viability of regional television as it has existed.

Under the Fraser Government a supplementary licence scheme was mooted to increase the choice in broadcasting services, while maintaining the regional operators' viability. A supplementary licence would permit existing regional television licensees in a single market to establish another commercial-television service, provided that it served the same region as the parent station. In an early submission to the SPS inquiry, members of the Regional Television Association (RTA) were united in their acceptance of one supplementary licence for each single station market, and they argued that 'in this circumstance it is very likely that the normal flow of commercial relationships between regional stations and potential SPS operators will bring about satisfactory introduction of SPS without the need for the introduction of additional regulations.'[22] Furthermore, they argued that supplementary licences would 'allow for the continuation and hopefully the flourishing of localism.'[23] On this basis they denied any need for further regulation. On mature reflection, however, they shifted their position, and in a final submission the RTA advised the Broadcasting Tribunal that its membership was now evenly divided on the issue of supplementary licences and SPS. Half of the RTA had opted for some kind of minimum regulation to protect their viability and independence.[24] The Minister was later to issue a press release which indicated that 31 of a possible 36 television stations had applied for supplementary licences, and that these would be subjected to a process of inquiry and decision making by the Australian Broadcasting Tribunal early in 1985.[25]

The basis of the regional commercial broadcasters' argument is for protection of their well established monopolies. Apart from shareholders of regional-television companies, few Australians will argue against offering additional channels to the 35 per cent of Australians who live in single commercial-channel areas. It was the intention of the broadcasting regulatory body in the 1960s, when commercial television was extended to provincial and country areas of Australia, that licensees would be drawn from business interests within the transmission area of the proposed station to enhance the concept of localism in program origination and selection. Yet, in a sample considered by the SPS inquiry, local programs only 'constituted between 2.5 and 13.7 per cent of total broadcast hours.'[26] A simultaneous inquiry into localism in Australian broadcasting found that estimates given by individual stations varied from approximately 9 per cent for GTW Geraldton, 12 per cent for TNT Launceston, 17 per cent for BCV Bendigo and GLG La Trobe Valley, to 31 per cent NBN Newcastle, for a week in September 1983.[27] Apart from some outstanding exceptions, the standard gesture of regional-television channels to localism is, for most channels, a twenty-minute local news service on most week nights. The notion of a channel with community participation in programming, similar to those which are standard offerings in cable television networks around the world, is not only not offered in Australian regional television, it is simply not understood.

The other consideration here is the phenomenal growth of advertising revenue attracted to commercial television in recent years. According to data compiled by the Australian Broadcasting Tribunal, in the financial year ended 30 June 1982, commercial-television revenues totalled $722 million, of which metropolitan stations earned $586 million and regional stations earned $136 million. Growth rates have been strong in regional television during the past decade. Total advertising spending on regional television grew eight times in the decade 1971-81, whilst metropolitan television revenues grew by six times.[28] In 1971, $19.3 million was spent on regional television which grew to $118 million by 1981, yet there was only a minimal rise in the number of broadcasting outlets.[29]

So, during this period of rapid revenue escalation the people throughout regional Australia have had no more choice in commercial television than when it was first introduced into Australia.

If ever an industry has thrived on economic protectionism, it is Australian regional television. A survey of company profitability by *Australian Business,* published in May 1984, showed that of the top Australian companies measured by return on shareholders' funds, a remarkable seven of the top 12 of all companies were regional-television organisations.[30] If satellite technology leads to greater choice for regional-television viewers, but at some cost to the profitability of existing operators, only the entrenched operators and their shareholders will grieve. Regional-television markets must be opened up, and preferably with programming sources which become much more diversified than under the established system.

The Tribunal's SPS report did not provide a neat list of problem-solving recommendations for the government, but turned the problems associated with satellite-delivered television firmly back on the policy makers. The report did not receive the commendation that it deserved for getting to the heart of the matter – the abnormally high levels of concentrated media ownership and control in Australia.

The Tribunal told the Government that if it wished to achieve stated Government objectives, then the fundamental issues of restructuring the broadcasting system had to be confronted before the problems associated with satellite broadcasting could be considered.

The ABT saw the fundamental problem as the present structure of control in Australian broadcasting. They pointed out that the Sydney and Melbourne television stations hold a position of dominant strength and influence within Australian broadcasting. The metropolitan networks, based in Melbourne and Sydney, really provide de facto control for all of Australian commercial-television stations by networking. The Tribunal pointed out that 'the metropolitan networks commission almost all entertainment programs produced in Australia; they control the sale of this material to the rest of the country's commercial channels; and they acquire a considerable proportion of overseas material which is broadcast on Australian commercial television.'[31] The corollary is the lack of opportunities for new television enterprises to enter the market. There was, in the view of the Tribunal, 'a lack of opportunities for expansion and growth in television by companies based in places such as Brisbane, Adelaide, Perth, Newcastle, Wollongong, Canberra, Hobart and in regional areas.' In practice, Melbourne and Sydney stations exert an economic ascendency over other commercial stations.[32]

The report made suggestions about how present commercial-broadcasting services could be extended to regional areas. Rejected was the proposal to introduce two supplementary licences on the grounds that it would only further increase the domination of Sydney-Melbourne networks. Also rejected was a proposal from the existing commercial-television networks based in Sydney and Melbourne, that people in remote areas should have access to the networks' satellite program services on the 30-watt transponders, in national beam. This would have made the three channels of commercial television available to remote areas, and apart from the cost of the equipment, the service would have been free. However, the Tribunal argued that this scheme would 'spell the end of regional services as Australia knows them.' Instead, an alternative kind of proposal, supported by Regional Television Australia (RTA), was proposed. This would be essentially a service consisting of a single channel of programming, licensed for each region, using a 30-watt transponder in a spot beam: a form of commercial HACBSS. This could be developed as a service to meet the needs of particular areas.

This suggestion was subsequently accepted by the Hawke

Government. In October 1984, Duffy announced that the 'network solution' was rejected in favour of a 'regional solution'. The Minister endorsed the proposal suggested by the organisation representing regional television licensees, Regional Television Australia (RTA), for a single channel of programming licensed for each of the four satellite footprints – essentially commercial HACBSS 2. Thus, argued the Minister, 'homesteads and remote communities currently without commercial television would be able to receive a regionalised commercial service via Aussat.'[32] Since the signals would be delivered via the 30 watt transponders in spot beams, the same dishes required to receive the ABC HACBSS 1 could also be used to receive the proposed commercial HACBSS 2.

The Minister's statement avoided the central problem outlined by the Australian Broadcasting Tribunal concerning the excessively high levels of concentration of ownership and control of Australian media, and the related economic dominance of the entrenched Sydney-Melbourne-based commercial-television networks. There are fundamental economic characteristics of the commercial-television industry in general, and satellite-delivered program distribution in particular, which are likely to further accentuate the tendency towards centralisation. The introduction of satellite-delivered program services from the networks is likely to further strengthen the pivotal position of the Sydney and Melbourne networks relative to other new members of the network, including new RTA licensees. The economics of satellite distribution are such that the network program suppliers are likely to generate additional revenue by the sale of as much national advertising as possible to be integrated with the satellite program feeds. The Australian commercial-television system is likely in the next decade to move towards the business structure which operates in US network television, with greater centralised distribution, production and decision making, both in programming and advertising.

In the absence of a detailed communication strategy based on social objectives, and with considerable nervousness about the viability of the satellite project, the Minister and his department have opted for a position of 'let's pass the buck to Aussat' and hope that ultimately the project will be a commercial success. This position manifests itself in ministerial statements that the government should not interfere with Aussat's commercial judgement about the allocation of transponders. Yet policy indecisiveness and delays on major issues have created serious commercial problems for Aussat Pty Ltd. One of the problems inherent in the ministerial announcement about preference for HACBSS 2 to RTA members was that these state-based groups, unformed in 1984, were obviously unable to file an expression of interest with Aussat as satellite users as a range of other prospective customers had done by May 1984. In fact, the Minister asked the ABT to conduct an inquiry 'to determine who would be the most appropriate applicants to be granted licences to provide regional commercial television services in four regional areas' and to report by April 1985 – a few months before the launch of Australia's first domestic satellite.[34] Meanwhile, the networks have filed their clear expression of interest with Aussat; companies in each of the Seven, Nine and Ten network groups intend to lease one 12 watt and one 30 watt transponder.[35] Aussat then must commercially juggle one 'regional solution' and the 'network solution' between the second and third satellites, and trust that it does not have too much underutilised capacity.

No matter how these business arrangements are ultimately resolved, the existence of the Aussat domestic satellites will confirm existing power bases in Australia's media, and reinforce the privileged access held by the few groups in commercial television to valuable public assets.

How Much Does a Domestic Satellite Cost?

There could be fewer more frustrating exercises in product accounting than the preparation of a balance sheet for a new commercial national service industry where the technology had a life span of only seven to eight years and for a company whose shareholders continually shelved policy decision so crucial to the development of marketing strategies. Financial data for Aussat Pty Ltd, Australia's domestic-satellite authority, has never been able to be accurately constructed.

There are many variables in estimating costs: administrative costs of the satellite authority, space-launch costs, construction costs of the satellites and earth stations, establishment costs of the command, tracking and telemetry stations and the costs of servicing the capital requirements. The inquiries which led up to the

Government decision to proceed showed how costs depended so much on fundamental political decisions - would the satellite authority be an independent statutory corporation, and if so how entrepreneurial was its role, what proportion of shareholders would be public and private, how much competition was desirable, or acceptable, with the existing telecommunications authority, and how would the satellite system be designed for prospective users?

Under the Fraser administration, Aussat Pty Ltd emerged as a fully commercial company which was expected to provide satellite services at acceptable tariffs in the marketplace, as well as generate dividend returns to its shareholders. And, of course, given the relatively short life-span of the satellites, the authority needed to strategically plan for replacement of the technologies and the establishment of a second-generation satellite program.

If cost estimates were difficult because of unknown variables, the question of revenue predictions was even more vague. How can a satellite authority really estimate demand for a service which has never existed before? How does it sell the capacity of the system when the managing body of shareholders - government - failed to decide which products can be marketed to whom? The reason why Aussat Pty Ltd had such an inordinate delay in the announcement of tariffs for satellite users was that they needed basic policy decisions to be made. The way in which the satellites could be used by broadcasters, and the extent to which Aussat would be allowed to compete with Telecom, were issues which remained unresolved for far too long, despite their crucial effect on the estimated financial viability of Australia's national satellite system.

The range of costs floated for Australia's domestic satellite project have been from initial 'ball park' estimates as low as $90 million to anti-satelliters 'flying opera house' projections as high as $980 million. The astonishing variations between groups on estimated costs is a major reason why people find it so difficult to come to a basic position on whether a national satellite is worthwhile or justified. A broad review of the history of overall expenditure as estimated by governments shows substantial cost escalation during the planning stages. In 1979 Tony Staley, elated by what he saw of the Canadian experience, claimed early on that it might be possible for Australia to have a satellite system of its own for $90 million.[36] His kind of broad speculation, based on the notion that everything in the electronics industry became cheaper in time, was not reflected in the Staley ministerial announcement to Parliament of 18 October 1979, that Australia was to proceed with its own domestic satellite.

He stressed that firm costings were dependent on the resolution of the overall system design, but the estimated cost of the satellite system was $210 million – broken down to $114 million for space-segment costs, $21 million for an earth segment of major earth stations and $75 million for earth stations funded by users. No administration costs were mentioned, and no decisions had been made at that stage on how capital costs would be financed.

A further revision of the costs was given in May 1982 by Ian Sinclair, as Minister for Communications, of $166 million for the cost of the Hughes contract for the three satellites and two ground-control stations, plus the costs of the launch vehicle and insurance. Aussat, it was announced, was to be financed by a mix of commercial loans and an estimated $75 million in equity contribution from the Commonwealth Government over the next few years. When asked outside the House for an all-in cost estimate of Government costs, including earth-station construction, he suggested that $650 million was a reasonable estimate.[37]

Two organisations with diametrically opposed views on the worth of a national satellite system – Aussat Pty Ltd, the system operator, and the ATEA, Telecom's technicians' union – have produced theoretical costs models which are useful to show the polarisation of the cost-benefits debate. Aussat's confidence that they are not in the game of marketing 'lemons' is shown in an unequivocal statement about their financial objectives, that 'extensive financial modelling has indicated that Aussat Pty Ltd should generate sufficient revenue during the life of the first generation of satellites to meet all of its operating costs and interest charges, pay tax, repay bridging finance, retain the real value of shareholders' funds and return reasonable dividends.'[38]

The ATEA, on the other hand, argued that real costs of the satellite system are much higher than generally predicted, and that Aussat's traffic predictions have been far too ambitious. According to the ATEA, 'Aussat's predictions of loadings, and hence the revenues which the system will generate, are unrealistic and take no account of the real market for telecommunications and broadcasting services in Australia.'[39] They argued that Aussat's prediction that both satellites will be fully loaded by 1988 are unrealistic, and that the system will be so under-utilised that there will be a substantial shortfall in revenues needed to meet operating costs. Somewhat boldly, they predicted that 'the shortfall between Aussat's costs and revenues will mean that by 1992 costs will have exceeded revenues by $459 million.'[40] According to Colin Cooper of the ATEA, all-in public costs of the national satellite project by 1992 will be $975 million.[41]

The on-costs to the public sector are substantial in terms of initial capital outlays, such as the cost of earth stations, and operating expenditure generated by the new services created to use the satellite. The three public bodies with guaranteed access to the first satellite – the ABC, Department of Aviation, and Telecom – all have new substantial financial commitments as a result of their involvement. In the case of the ABC, apart from the transponder-leasing costs of an estimated $25 million annually for its preferred 9.5 transponders (which must be offset against reductions paid to Telecom) there is an extensive earth-station program. Minister Duffy announced for the 1984-85 budget that earth stations were to be installed at all 41 of the ABC's present and proposed regional studios at an estimated cost of $15.05 million over a six-year period.[42] The Department of Aviation was given approval by the Fraser Government in August 1982 to use the domestic satellite from the start of its commercial operation – 100 ground stations and additional estimated costs of $31 million over six years.[43]

Telecom has a double cost bind. Not only does it have an initial capital outlay for satellite equipment such as ground stations, but it faces loss of revenue particularly as a result of the reduction of use of television relay carried by its microwave bearers. In 1983 the ATEA argued that competition in the inter-capital city voice and data markets 'will result in the loss of $300 million per annum by Telecom.'[44] The question of on-costs to the public sector was consistently avoided in Government statements about the cost benefits of Australia's domestic satellite. Even after a change of Government in 1983, when the opportunity arose for a cost review, the decision to proceed with the satellites by a Labor Government was apparently made without any detailed costings of the project.[45] Neither the Labor Cabinet nor Caucus paid much attention to financial analysis of this immensely expensive project.

The cost-effective satellite argument, cleverly dominated by lobbyist union the ATEA, centred on Aussat's financial vulnerability and perceived income losses to Telecom. An unknown variable is telecommunication growth, or the catalyst effect on the telecommunications market that the existence of a domestic-satellite system might have for new information services. Aussat management have pointed to the growth analogy that no one ever anticipated the phenomenal acceptability of the Intelsat satellites. In early 1984, when companies and organisations were asked to file their prospective level of usage of the satellite system, Aussat management announced that the level of demand was in excess of what had been anticipated.[46] Subsequently in October 1984 the Ministers for Communication and Defence Support announced that European company Arianspace had won a $24 million contract to launch Australia's third domestic satellite. The Ministers said that advice from Aussat had indicated the launching of Australia's third satellite might need to be advanced to mid-1986 because of the high demand already registered for the first two satellites.[47]

Aussat's commercial stability will be heavily dependent on guaranteed revenues from Australia's major television organisations. In the years leading up to the launch, Aussat staff members admitted that broadcasting revenues were absolutely crucial to the ultimate financial success of the project. Their estimates of the relative significance of broadcasting revenues to total Aussat revenues varied from 50 to 80 per cent.[48] There are several areas of doubt about the viability of Aussat for television corporations. The Australian Broadcasting Tribunal has expressed its view that it is not clear 'that the distribution of television programs by satellite will necessarily be a financially viable proposition for Satellite Program Service providers in either the short, or medium terms.' At the ABT hearings, evidence given by staff of Melbourne's Channel Seven argued that both the satellite and the supplementary licence scheme were 'very expensive problems looking for revenue solutions'; Sydney's Channel Seven representative explained that the satellites 'will add convenience to program distribution but at a 300 per cent increase in cost to the networks'; the dual Channel Nine submission stated that 'the cost effectiveness of SPS . . . remains to be seen.'[50] The ABT also expressed viability doubts about the proposed Regional Television Australia's HACBSS 2 solution, particularly if it is a free service.[51]

Moreover, Aussat's biggest public-service television customer, the ABC, is precariously dependent for income on the government of the day, with fluid annual budget priorities.

Aussat Pty Ltd, which has some measure of vulnerability with assured long-term television revenues, is not able to offer a public switched telephone and data service. The *Satellite Communications Act* 1984 confirmed Telecom as the monopoly public switched telecommunications provider for Australia. Where Aussat wishes to compete with Telecom in offering private leased networks, the satellite authority will find a formidable rival in the incumbent telecommunications giant. In areas of data transmission, Aussat will compete with Telecom. Telecom has been well aware of the potential in the data-transmission market since it first offered data services - Datel in 1970. Demand has grown steadily and the development of Austpac and DDN, Digital Data Network, have been hastened in the 1980s. By virtue of its established corporate strength, Telecom could adjust its rates to maximise advantages for particular services, such as data, as it chooses. Telecom could price its tariffs for data services so attractively that Aussat will battle to find customers in areas of data telecommunications. Rarely publicised has been the view expressed by shadow Communications Minister Bruce Lloyd in debating the *Satellite Communications Bill* 1984 that the Telecom elephant is being protected from the Aussat mouse.[52]

Australia joins the space age of domestic telecommunications in 1985. Under the Fraser administration the hidden agenda for its justification was the development of a second national telecommunications network, though it was sold politically on the basis of improved services to the outback. There is little doubt that for isolated rural communities the top priority for communication facilities is a telephone service, yet satellite technology is an immensely expensive way of providing outback telephony services. The rationale for Australia's domestic satellite system primarily emerged as an alternative television transmission-delivery system and for providing an ABC and commercial television service to isolated communities via their own small earth stations.

Aussat will bring other changes to television-distribution patterns. New broadcasting outlets are justified, especially in established regional-television markets where more channels are needed. The continued healthy buoyancy of the commercial-television industry, even during a period of economic recession, would indicate that the industry across the board can accommodate new players. Little consideration has been given, in both public-service and private television organisations, to the prospects of programming innovations which might accompany satellite technology, and only limited attention to the cultural effects of outback broadcasting.

The shape of telecommunications in Australia in the future is likely to be fashioned primarily by economic circumstances. Given Australia's substantial national public debt, it is conceivable that there will be a future era of tough monetarist policies. Allied with strict monetarism will come a revival of the call for the extensive privatisation of telecommunications. Telecom's monopoly could again be substantially threatened, and Aussat invited to run competitive services as a more privatised telecommunications institution. The future of the domestic satellite system, and Australia's telecommunications network, is inexorably linked to the nation's overall economic situation and to our choice of who are political masters of communications policy in Canberra.

10 New Networks: Culture, Content and Cables

Information transfer within and between organisations, and within and between nations, is one of the fastest growing businesses of the mid 1980s. Banks and insurance companies, financiers, gambling authorities, hotel and travel organisations, airlines and transport companies, computer-service industries and others, are using fast and efficient information operations to link their dispersed member enterprises. Information technologies, particularly computers and word processors, are making contemporary commerce more dependent on complex information networks. An OEDC report observes:

> The production, transmission and processing of the most varied information will be at the heart of economic activity and social life . . . Through its links with data processing and telecommunications, the electronics complex during the next quarter of a century will be the main pole around which the productive structures of the advanced industrial societies will be reorganised.[1]

Information as a commodity is part of a new phase of capital accumulation, made more efficient by the new electronic means of transfer, distribution and storage.

Electronic Culture

New communications infra-structures are being built in advanced economies, signified by labels such as 'electronic highways' or the 'wired society'. If Australia ever reaches fully wired status, mid-1980s indications are that this will come about much more gradually and hesitantly than the electronic utopians would have had us believe. While new technologies not subject to regulation – such as computers, word processors and videocassette recorders –

have received a rapid acceptance in the Australian marketplace, regulated technologies for Australia, such as cable and subscription television, may never be introduced. However, for other regulated new technologies, 1985 heralds the introduction of a national public videotex system, Telecom's Viatel service, and domestic-satellite transmissions via Aussat.

In advanced industrial societies, computers, word processors and telecommunications equipment have become the key components of 'the automated office'. Simultaneously, there has been a trend, albeit slower, towards 'the electronic home', based on new products services and delivery systems, such as cable and pay television, videocassette recorders, home computers, videotex and teletext, and the videodisc. Market acceptance of the automated office and the electronic home differs from country to country. In advanced economies the rate of technological development is so fast that it is conceivable that office and home may converge in the future in universally interconnected electronic-communications systems based on signals such as broadcast, cable, microwave, satellites and optical fibres, delivered to both.

Paradoxically, the age of information abundance is also an age of increased social deprivation. As social and economic practices in Australia become more dependent on new forms of electronic information processing, greater information inequity is developing.

Several issues emerge. Will this new electronic estate provide greater abundance, with greater choice available to all? If information is now a primary commodity, will information be widely available to all, or merely to those who understand information systems, and who can afford the new information services? New media demand a new cycle of investments - the audience has to buy or rent the videocassette recorder, the pay-television decoder or the videotex black box. It is only for those who can afford it. Moreover, new communications technologies such as cable or subscription television, and videotex, work on a direct user-pay principle. Hence, the extent to which consumers of the electronic home opt to take up the new broadcasting systems depends on household income, particularly so with expensive interactive services offered by cable-television operators such as home security systems.

In the early development of the British videotex system, there was only substantial take-up in the business sector and by affluent, already well informed households. What is emerging is a widening gap between economic classes, and an increased cleavage between an information-rich elite, with an understanding of and access to data bases, and a growing class of information poor.

New electronic networks, where computerised databanks communicate with each other, have given a new context to an old debate about personal rights of privacy. New technology has exponentially increased capacity to store information, whilst the convergence of computers with telecommunications has created greater mobility of data, within and between national boundaries. Vast quantities of personal information are now stored, virtually indefinitely, about almost every member of society – a personal medical, financial, legal, educational and employment history. Crosslinkages between computer systems, storing different data about the same person, makes feasible the construction of comprehensive personal dossiers. Electronic record keeping of many routine activities, from bank deposits to medical visits, has produced information efficiency, but the personal security risk involved has been heightened.

In New South Wales, a specialist body, the NSW Privacy Committee, was established in the mid 1970s to investigate general complaints about privacy invasion. The most pertinent examples of complaints over a five-year period included refusals to grant credit due to mistaken identity; failures by retailing organisations to update their records leading to incorrect listing of clients as defaulting debtors; and disclosure of an employee's debt to an employer. Obviously there has been the development of computerised blacklists.

In a major report on privacy, the Law Reform Commission points out that 'the introduction of the computer into personal record keeping provides the incentive, and the means, to invade the information privacy interests of data subjects.'[2] The potential problems are compounded by the inaccessibility of the information-allowing invasion of privacy to occur without its ever being discovered by the victim. The Law Reform Commission's privacy report argued that if a person wished to intrude into the private life of another, or use private information about another, for a purpose of which the subject would remain unaware, and for which they probably would never have given consent, the subject was now virtually powerless to protect himself, and there are few legal protections in Australia upon which to call.[3]

New television-distribution systems such as cable television also have inherent problems of potential invasion of privacy. Some cable-television systems in the USA permit viewers at home to respond to questions asked on screen by keying in their response, and there are considerable inter-active possibilities such as teleshopping, telebanking, and fire and burglary systems. In some systems, billing is on the basis of individual programs viewed, where the computer scans all homes connected to the system every few seconds to record which programs are viewed. Obviously, dossiers of some behavioural habits can be built on those who subscribe to the system. These could cover purchase of consumer goods, the position taken by viewers on survey questions, extent of viewing of pornographic films, subscriptions taken out, times that owners left homes or turned off alarm systems, health data involving special alerts, and contributions to political or charitable bodies.

All of this data can be collected through an inter-active cable television system and stored in a central computer controlled by the cable operator. Some cable companies in the USA have developed their own privacy rules as a voluntary regulation of information practices. For example, the Warner Amex code, issued in October 1981, assures its customers that no individual information concerning viewing or responses is to be developed 'unless the subscriber has been advised in advance and given an adequate opportunity not to participate.'[4] In some states a cable-privacy statute has been enacted, though no national code of protection is ever likely to emerge in the USA.

Australian law has not developed a concept of privacy and the common law has not developed a general right to privacy. The Law Reform Commission has argued for the establishment of a statutory body with the specific function of advocating privacy interests. They argue that such a body should not be limited to the protection of privacy merely as a policing authority, but should be able to consider privacy problems in the broadest context of a society undergoing rapid change.

The report proposes limits on the collection and use of personal information and its disclosure. In their view, the Human Rights Commission should be enlarged to include the Privacy Commissioner and assume special responsibility for privacy protection. The Commission also sees an important role for self-regulation, and, where appropriate, significant coercive powers in the area of access to and correction of data records. The thorough report of the Australian Law Reform Commission on privacy, seven years in the making, has received all too little attention by the Australian Parliament and the community at large. Unless policy is formulated in this area, large organisations, including governments, will be able to use the new information technology to serve their own interests at the expense of serious potential abuse of the privacy of individuals.

The growing use of new communications technologies also raises many other issues of change for established media institutions and practices. What are the implications for the 'old' media of 'new' media? Gurus such as Alvin Toffler have referred to the 'demassification' of media brought about by the plethora of new distribution systems and fragmentation of the 'mass mind'.[5] Similarly, Anthony Smith, author of *Goodbye Gutenberg,* claims that new media 'provide opportunities for individuals to step out of the mass homogenised audiences of newspapers, radio and television and take a more active role in the process by which knowledge and entertainment are transmitted through society.'[6] For Ithiel de Sola Pool, American author of *Technologies of Freedom,* electronic media are coming to be so dispersed in use and abundant in supply that they 'allow for more knowledge, easier access, and freer speech than were ever enjoyed before.'[7]

In a great deal of writing about new media, the prospect emerges of the transformation from 'homogenised mass media' into new media which serve many different publics with much more participation from consumers. These prospects are exciting, yet in the mid 1980s they remain merely promises. So many writers seem to assume that these new communications technologies are on the verge of ubiquity. Even in the United States technological hothouse, the notion of the electronic home for new media has a long way to go: cable television does have a market penetration of 40 per cent, but separate radiated subscription television services have only two million American subscribers. Fewer than 10 per cent of American homes have home computers and videocassette recorders. Videotex and teletext systems have been barely developed.[8]

The emergence of new competitive electronic delivery systems has led to renewed predictions of the imminent departure of print media, especially daily newspapers. There have been odd closures, some surprising, such as *The Washington Star,* but the broad pattern of daily newspapers is generally one of static or slightly declining circulations around the world. In Australia during the past decade, only the *Age* and the *Australian Financial Review* have shown marginal circulation growth. The newspaper-industry share of total advertising revenue in Australia has dropped significantly in the last decade, with the sharpest drop for metropolitan dailies, where the afternoon press appears to be losing its share of revenues most rapidly.[9]

The two extremes of the daily-newspaper market appear to be the least threatened. Upmarket papers, the 'quality' press, with their appeal to the high-income professional-business strata and with a strong classified-advertising base are well placed. As well, the downmarket 'tits and bums' papers, with large doses of sport, trivia and pictures, have held their appeal, and they are likely to keep their major advertisers and maintain circulation, with occasional reader bingo blackmails. Newspapers which have traditionally held the middle ground tend to be suffering substantial circulation loss, notably Melbourne's *Herald,* whose daily sales have dropped by about 200,000 copies since the early 1970s. What tends to be forgotten in discussions about circulation loss and advertising-revenue fall-offs for daily newspapers is that, despite their problems, they have still provided the cash machine for major groups, such as News Corporation Ltd and John Fairfax Ltd, to expand their media empires.

Rupert Murdoch has approached newspaper purchase and developments in terms of the way markets have become polarised: in the UK the upmarket *Times* and the downmarket *Sun,* and here their counterparts such as the *Australian* and the *News* respectively. These kinds of newspaper-market trends need to be explained in terms of changing community factors, but it is likely that any further competition from electronic media will continue to damage most middle-ground afternoon-daily newspapers.

The newspaper industry has experienced substantial changes as the result of the introduction of new technology within existing plants. Most Australian daily newspapers are now completely set by computerised typesetting methods and some are now direct printed. In many organisations journalists have had their typewriters replaced by video display terminals (VDTs) which enable stories to be readily altered, instantly called up, and indefinitely stored. An international correspondent of an Australian newspaper can file copy on a VDT in Washington, which can then be transmitted around the world at a rate of 1200 words a minute, edited in Melbourne or Sydney, and sent straight for photocomposition. Classified advertisements are now increasingly handled by telephone operators who put the advertisement directly onto a VDT and thereby instantly into the computer system.

These new methods of production are less expensive than the old, in plant as well as labour. Moreover, because less plant space is required, newspaper publishers have been able to sell unwanted real estate.[10] For Harry Gordon of the Herald and Weekly Times group, the new technology brings 'considerable improvements in efficiency' and the chance for the newspaper industry 'to beat its massive cost problems.'[11] The human face of such change, and some agonising industrial-relations problems, has been well described elsewhere.[12] For Greg Taylor of the *Age,* in the long run 'the new technology will lead to better newspapers because of the increased speed, lower labour costs and the capacity of newspapers to devote more of their financial resources and energies to collecting the news.'[13]

Marshall McLuhan, the 1960s oracle of the electronic age, ironically became a best selling author of a book predicting the death of print. Newspapers, too, are hardly ever likely to disappear: their cheapness, convenience, portability and permanence will ensure their continuation. There will, however, be a strong trend towards electronic publishing, where material is added to a central data store and made available to those who have the techniques and means to access it. Using computer logic, an enormous capacity of information can now be transmitted and stored electronically, with a flexibility and cheapness unmatched by traditional print methods. Electronic storage is cheaper for bulk information than using filing cabinets, and libraries have come to widely adopt computerised storage of data to save space.

The development of electronic publishing will eventually lead to profound changes in the publishing process. Anthony Smith has argued that the new communications technologies will contribute towards the erosion, but not abolition, of the Gutenberg principle. We are liable to witness, he argues, 'a rapid erosion of the settled notion that information is naturally multiplied in physical copies until the number of copies approximates the number of those wishing to receive it.'[14]

The process of broadcasting undermined the Gutenberg principle in that the mass multiplication of physical materials, as in book and newspaper publishing, was absent. According to Smith, we are moving towards a new condition where 'an individual can choose electronic dissemination of a single item that was itself chosen from a vast or total store of video products.'[15] This condition is closest in the field of data, with on-demand publishing from a demand repository using a microcomputer, or through a videotex system. The *New York Times* has an established on-line library database for access via dedicated terminals. In the Australian newspaper industry, AAP and the *Australian Financial Review* have developed specialised computer-access information services.

Newspapers only publish a small fraction of the total information possessed, and it is conceivable that they could eventually produce a product which is a separate newspaper for each reader. Here each reader would be provided with selected articles as a representation of individual reader interests, for a price. These kinds of possibilities again beg the question of whether we are witnessing the end of the penny press associated with mass literacy, and the beginning of a new technological era designed for an elitist group of electronic aristocrats.

Videotex is an important manifestation of electronic publishing. Videotex is an information-retrieval system which uses a central computer to store and process data, the existing telephone network to distribute the data, and modified television sets (or microcomputers) to display the pages of information selected for viewing. The data is not broadcast over the air, but is sent as an audio signal over the public switched telephone lines to an adaptor connected to an existing or modified television set.

The system is a two-way information flow in that the user sends commands to the computer data bank, which responds by returning the appropriate information to the television screen. The information that can be made available includes business data, such as stock-market quotes, exchange rates, company reports, commodity prices, world economic indicators or general consumer information such as weather reports, travel and holiday data, news headlines and restaurant, theatre and cinema guides.

There is really no limit to the number of pages which a videotex system can offer and, in its most advanced form, it could bring a massive reference library or teleshopping and telebanking into the home.

Videotex systems have been largely developed by state public-telecommunications corporations around the world – it was first developed by British Telecom as Prestel, followed as Teletel and Antiope by the French PTT, and later fostered by the Canadian Department of Communications as Telidon.

The prime role of Telecom in offering such an ambitious service has been the subject of political football during the past few years. In January 1981, the then acting chairman of Telecom indicated to his Minister that Telecom was keen to offer a videotex service similar to Prestel. He believed that such a system could develop into a valuable national information resource, and that it would be of advantage to both Telecom and the private sector. However, Communications Minister, Ian Sinclair, deemed that Telecom should not proceed, and that private enterprise only should be permitted to operate videotex systems. This was at a time when the Fraser government was keenly exploring the greater privatisation of telecommunications and Sinclair's hunch was that videotex might be best left exclusively to the private sector.

A number of small-scale videotex systems developed in the private sector, such as Thomas Cook's 55-terminal national travel-industry network, General Motors Holden's stock-locator system, the Qantas base for accurate information on airline schedules, and the ANZ banking group's corporate treasurer and business managers' videotex service, Anztell. The rural sector has seen the most competitive market push, with a three-way marketing battle in rural commercial videotex systems between the Elders pastoral group, Grassroots Australia and Agridata Australia Pty Ltd. Some videotex systems are for internal house use only, while others are publicly accessible data bases. Inevitably, though, without the central involvement of the national common-carrier telephone authority, Telecom, videotex services were bound to remain fragmented, underdeveloped and expensive. With the election of a Labor government in March 1983, together with the arrival of pro-Telecom Communications Minister, Michael Duffy, Telecom's own videotex system, Viatel, was approved for introduction throughout Australia in February 1985.

Telecom's prime role in Viatel is to be the common carrier of information – Telecom undertakes to operate and maintain the system, but not to provide the bulk of the information offered to customers. The information to be retrieved is provided by independent service providers who offer and update the data and lease pages to consumers. The system providers can be private companies and government departments or agencies, and gateway facilities will allow private computers to be connected to the public videotex service. In no country to date has a public videotex system achieved substantial market penetration, still less profitability.

The introduction of the Australian national videotex system, Viatel, has the circular marketing problem of trying to attract consumers with limited services and trying to attract information providers to the limited market. The early days of British Prestel's videotex service, offering mainly weather, news and sports results, failed to gain wide consumer demand. Viatel's initial development phase would seem to be catering more for the business market with, hopefully, customers willing to pay for more specialised data. Yet the long-term success for Viatel is its ability to develop widespread community data services.

Videotex has several inherent advantages as a medium. It can be updated more rapidly than any printed material; videotex information can provide a vast array of stored data, at the user's convenience and time of need; information, not available through newspapers, radio and television, can be made available to specialised users; and the users only pay for the information they require. Its principal disadvantages are that it is more difficult to assimilate continuous text through using videotex than through the printed word and, of course, in videotex there is an absence of hard copy.

Ultimately, a videotex system will succeed in Australia only if people want the information provided, and are prepared to pay the cost. The Australian experiment will be fascinating to watch as it helps answer the international question – is videotex a technology in search of a market?

Videotex is perceived by some as a long-term threat to the revenue base of newspapers. Information carried now in classified advertisements can be instantly updated by consumers using a videotex terminal. The stakes are high in such competition from videotex; the *Age,* for instance, carries 1.5 million classified advertisements annually which generate $50 million in revenue.[16] Videotex can target advertising down to individual households: Ken Shard, marketing director of London's *Daily Telegraph*, has warned that videotex could expect to capture advertising from newspapers in areas of 'used car sales, house buying, job hunting and the entertainment, travel and shopping sectors where ordering and buying could be done at the same time.'[17]

Such a threat poses major strategic corporate-planning considerations. Extreme options have been outlined by Howard Anderson, president of the Yankee Publishing group: either 'use newspapers as cash cows and milk them for all they are worth, before closing them down and re-allocating the resources,' or, do not worry because newspaper companies 'have got the data bases, the reputations, the savvy, and they will be the determining factors.'[18] There are clear signs in Australia of newspaper organisations building strong software bases for the coming electronic age.

The long-term success of electronic publishing will be affected by the degree of acceptability of personal computers. Their development represents a remarkable technical achievement in that small personal computers can now offer about the same computing power as a large mainframe of the 1960s or a minicomputer of the early 1970s. The first personal computer appeared in 1975, and by 1980 about 24 firms had sold 750,000 units, for about $1.8 billion.[19] Two years later another 20 or more companies, including IBM, had entered the p.c. market, and aggregate world sales were estimated at $6 billion.[20] There has been no absence of industry predictions for a multi-billion dollar growth, yet by 1984 there were indications of a levelling of demand. Signs of an industry shake-out appeared with an original pacesetter, Texas Instruments, withdrawing from the market, and with Warner's Atari and Osborne in some trouble.

A personal computer is a small computer based on a microprocessor and hence the usual generic label of microcomputer. The drive for initial sales tapped the market of those who wanted to use a personal computer for video games. Yet these computers have the basic attributes of computing – manipulating thousands of numbers per second, and the ability to store, process and retrieve huge amounts of data. There has been a strong demand from those involved in writing in some way, be they teachers and students, lawyers, secretaries, accountants, researchers or business people. Software programs are available for purposes such as keeping household-finance records or developing an inventory of some kind, or for income-tax records. It is also possible to use simple programming language to type up essays, articles or reports, call them up on screen, get a printout, and store them within the system. The market has not only seen a demand from independent professional workers but, increasingly, large organisations are developing individual computer work stations to process their information. At its most advanced stage, the automated office has individual computers linked to each other, and to large central computer facilities through local area networks.

The lobbyists' toy has now become a new technological marketing frontier for big corporations: the major American brand names are Apple, Commodore, Hewlett Packard, IBM, Osborne and Tandy, with leading Japanese exporters such as Hitachi, NEC and others. As examined earlier, Australia is a colony in terms of computer hardware, though indigenous companies such as AWA and Hartley are in the personal-computer field. There are virtually no industry worldwide standards in this market, so novices find a bewildering range of peripherals – display screens, disc drives and printers – and the software from one system is generally incompatible with that of another system.

The growing dominance of IBM in the market will probably mean that eventually it will provide the industry standard. In an unregulated market, such as the computer field, a successful corporation can establish de facto market control for the future.

Though reliable data is hard to come by, it has been estimated that IBM shipped 670,000 personal computers in 1983 and more than three times that number in 1984.[21] In Australia, market estimates for the first full year of its personal-computer sales was 4000 IBM units, later followed by market leadership. IBM has developed its personal computers with highly derivative products, such as the disc drive supplied by the Tandy Corporation, and with reliance on software from outside sources. Not surprisingly, most programming organisations are now designing software for IBM personal computers.

The marketing methods for personal computers have often played on fears of failure in school and life, using encyclopedia-sales techniques in a high-tech disguise. Personal computers are still a comparatively expensive electronics product, predominantly used and understood by professionals. It remains problematic as to whether they will become an easily affordable component of the electronic home and whether a range of new home-based occupations will ever evolve around the notion of an electronic cottage.

The proliferation of new communications technologies also presents the advertising industry with a series of new challenges. New media are more specialised, reaching fragmented audiences, and as a consequence advertisers are finding that media consumers are more spread than ever before. In many ways advertisers were well served by the 'old' media, and particularly commercial television. Though they have felt aggrieved by the outrageous price hikes foisted upon them by commercial-television management in recent years, it remains a maxim of advertising that commercial television is the most effective media for selling products. The protected closed markets of commercial television have well served both the television industry and the major advertisers. The tendency towards audience fragmentation will pose problems for advertisers, in that media buying will become more problematic and the cost of finding consumers through advertising is unlikely to reduce.

One of the major trends in the advertising industry as a whole, the growth of direct selling, is likely to gain a fillip from the acceptance of new communication technologies. Direct-marketing techniques include mail ordering, tear-out coupons, catalogue ordering, and phoning the flashing number on the television screen. In recent years the direct-marketing divisions of major advertising agencies are all claiming substantial increases in billings. In highly competitive areas, computer-based response techniques, such as electronic funds transfer, represent an advanced phase of direct consumer response. If interactive systems eventually expand widely, such as those available through videotex or cable-television systems, advertisers will have an effective pipeline to consumers. Computer-based inter-active systems have the potential, (and the privacy risk for consumers) to provide advertisers with a data base about consumer habits and their socio-economic status.

In terms of buying media space, new communications technologies are likely to present advertisers with polarised choices. What may emerge is that commercial television will remain as the only mass-advertising medium for at least the next decade, and that other media, such as radio, newspapers, magazines and cable television, will target on more defined but narrowed audiences. The introduction of Australia's domestic satellite is likely to strengthen the strategic position of television through networking, and major advertisers will be attracted to the national distribution of satellite programs. National advertisers will find that network television gains an increased attraction in that satellite delivery presents bigger, more centralised markets. On the other hand, advertisers who opt for new media are likely to find specialised markets at affordable rates, yet agonise over whether their choices are cost effective.

An issue of concern to advertisers is that the new media provide ways of avoiding their advertisements. In this context, the video-cassette-recorder revolution has been perceived as a threat to the advertising effectiveness of television. Of all the new media technologies, the VCR in Australia may be the only item of electronics equipment since television itself to eventually achieve mass-market acceptability. Video has many attractive attributes – it requires no complex technical infra-structure, is a simple means of program dissemination and the hardware cost has become progressively cheaper. Video watching has tapped latent dissatisfaction with commercial television, and shown that demands for media diversity represent more than mere obscure academic interest. Video offers selective program recording, time shifting, and the viewing of contemporary films at readily affordable prices.

According to industry estimates, video-hardware sales in Australia have rocketed from 26,000 units in 1979, to 150,000 in 1981, to 550,000 by 1982, to over one million by the end of 1984.[22] Estimates vary, but the figure for VCR penetration by the end of 1984 was generally in the 25-30 per cent range. With an estimated 3000 video libraries throughout Australia, video is now estimated to be a $250 million industry.[23] The video revolution has been a goldmine for Japanese-manufactured technology and American movie-programming corporations. The bulk of VCRs are manufactured in Japan, with Sony and Matshushita as leading corporations, and Japan export sales run at 10 million machines annually. American companies tend to dominate the pre-recorded software movie packages, chiefly from the studios of 20th Century Fox, Paramount, Walt Disney, Warner Brothers and United Artists. It has also, of course, given Australian film producers the opportunity to release their products through video-distribution outlets, but with mixed success.

Research conducted by the Television Bureau of Advertising, the television ratings-research company, McNair Anderson, and advertising agency, George Patterson Pty Ltd, has concluded that VCRs do not present a threat to traditional television-viewing patterns.[24] Yet others within the industry have argued that the estimated three to seven-and-a-half hours a week of video viewing indicated by these surveys, without commercials, must concern all advertisers.[25] Advertisers also fear that when videos are used to record and play back programs, viewers are 'zapping' out the advertisements with fast-forward buttons. The process of record-and-replay means that advertisers are no longer sure that their prime-time commercial will be seen by the prime-time viewer.

Cable and Pay Television

One of the major new electronic-systems choices for new media in Australia is whether to introduce cable television (CTV) or pay television as radiated subscription television (RSTV), or both. Since Australia has no experience with major alternatives to conventional over-the-air broadcasting, the American model of cable systems usefully highlights some of the major issues to be addressed.

The USA has the greatest number of cable systems of any country – more homes are connected for cable television in the USA than in Europe, Canada and Japan combined. Predictably, the ownership of cable systems in the USA has come to be dominated by big operators.

The largest systems operators now tend to be subsidiaries of major media conglomerates: ATC, the second-largest American cable operator, is owned by Time Inc, whose interests are in magazine and book publishing; Group W Cable is owned by Westinghouse, a major broadcaster and the fifth-largest USA electronics manufacturer; Cox Communications is the tenth-largest USA publishing house; Times Mirror is another major publisher and station owner. The Hearst Group, the largest publishing house in the USA, moved into cable systems in 1980 and is now planning to provide extensive data-base information systems. American Express joined with Warner Communications in 1980 to create a major multiple-system operator and the programming company Warner Amex. In 1978, Time Inc created the first major pay-television national network in the USA, the now lucrative Home Box Office. In the USA, the entrenched commercial-television networks have found themselves surrounded by a new league of formidable television-systems competitors.

Cable television is an alternative method of television delivery, with transmission via co-axial cables, overhead or underground. Frequency space is limited with over-the-air television transmission – there are no more VHF channels available to Melbourne or Sydney – but signal delivery through cable systems could offer dozens of additional channels. Cable television involves direct billing to the household, usually on a monthly basis, for the program services offered. There are two forms of payment, a flat fee for a fixed number of extra channels, called basic cable, and additional tiers of payments for the 'premium' channels, usually feature films, called pay or subscription television. In the USA a basic cable package, for $12 a month, generally offers about 20-30 channels, as a mix of distant television stations, local over-the-air channels, local community-access channels, and satellite-delivered programs of news, sport and films.

Many channels narrowcast to specific audiences: subscribers may be offered 24 hours a day of SIN, the Spanish International Network of Spanish-language entertainment programs, the Trinity Broadcasting Network, CHN, the Cable Health Network, and MTV, a never-off-the-air rock-music channel. As well, for most of the new cable systems, over 90 per cent of subscribers take at least one pay-television channel in addition to the basic cable package – generally a new-release movie subscription channel for about $10 a month.

Newer cable-television systems in the USA offer a range of additional non-broadcast services. The best known of these is the two-way interactive Qube system in Columbus, Ohio, where viewers can react to the programs by keying in a response to designated multiple-choice questions. A cliché of the companies is that these facilities will change television from an entertainment medium to a communications medium. The broad range of interactive services also includes security services, information-retrieval services, telebanking, teleshopping and video games.

Inevitably, claim the cable operators, it will lead to telepolitics – vote for your President from home! At present, home shopping and home banking remain in an experimental stage, but the burglary, fire and medical alarm systems are the most advanced inter-active cable systems, for those who can afford them.

Radiated Subscription Television services are transmitted over the air like a standard television service, except that the program signal is in a coded or scrambled form. To secure a viewable signal the RSTV customer pays for a decoder which unscrambles the signal. The staple diet of RSTV programming is movies, uncut and uninterrupted by commercials. Costs to the consumer in the USA currently average about $20 a month for RSTV, plus an installation charge. By comparison with cable television in the USA, RSTV is underdeveloped, currently with about two million subscribers throughout the USA. RSTV operators have introduced the 'pay per view' concept, where only those subscribers who have paid an additional fee are able to receive the telecast. This has been used for boxing events, such as the heavyweight-championship fight between Holmes and Cooney, and live concerts by the Rolling Stones. The potential here has not escaped the Hollywood moguls: MGM/UA Entertainment co-chairman, Frank Rothman, 'When 20 million homes are equipped to receive pay per view, we could charge $5 a home for Rocky III, split the profits, and make $50 million in a single night.'[26] Though film-studio management has welcomed pay per view because of its box-office basis, they are concerned that pay-television distributors will exert great industry power if the concept develops widely. In the long term it poses a serious threat to established practices in cinema exhibition.

Cable television was originally designed for viewers who could not get pictures with conventional antennas, or who had limited off-air reception. Its beginnings go back to 1948 when it was known as community antenna television – CATV – in areas where terrain or distance from stations prevented television reception. A master antenna system was erected at a site where a good-quality signal was available. These signals were then relayed to the community for distribution to individual subscribers via underground co-axial cable or overhead wire. Since many subscribers shared the same system, receiving equipment of high quality was installed, providing clearer pictures than home antennas, particularly for colour television. Such systems offered not only a better quality picture, but additional channels, FM radio and other services by cable.

The cable-television industry in the USA remained small and underdeveloped; in 1950 only 70 cable-television systems were in operation in the USA, serving a mere 14,000 subscriber homes. Thirty years later some pundits now claim that the big networks have lost the battle to cable television. At the end of 1983 there were 83 million television households in the USA and over 34 million cable subscribers – a 40 per cent cable penetration of television households.[27] An extension of this to 60 to 70 per cent of American households by 1990 is widely predicted by cable-marketing researchers and financial analysts.

Where cable-television systems developed in the USA from rural to urban areas, the tendency was reversed in Canada. Canadian cable television began in the early 1950s – before Canadian television itself. The central purpose was to receive American television channels from cities near the Canadian border, out of reach of a conventional home antenna. By 1952 cable companies were operating in Montreal, Vancouver and Ontario, offering cable subscribers local and distant Canadian signals as well as American programs. Large cable companies have existed for years near the US/Canadian border, in Montreal, Toronto, Winnipeg, Vancouver and Quebec, but cable has now spread to outback Canadian regions, and only a handful of small communities are yet to be licensed. In 1983, 505 cable systems served 4.6 million Canadian subscribers, representing almost 60 per cent of television households, or a remarkable 75 per cent of Canadian homes passed.

Some important aspects of the phenomenal growth of the American cable-television industry are worth noting. Cable companies scouted for as many program inputs as possible and allied technological developments made possible the proliferation of 'distant signals' into the communities they served. Cable companies imported 'distant signals' via microwave from the closest city receiving conventional television. The biggest breakthrough came in the mid 1970s when satellite communications greatly expanded the program capacities of the cable systems. With satellites, a multiplicity of signals could be beamed out, to be picked up by earth stations and then distributed to viewers through the cables.

Television-production companies blossomed in many parts of America – super stations for movies, news, sport and drama. The effect of the development of domestic satellite communications on the growth of the American and Canadian cable-television industries cannot be overestimated.

Many cable companies who had struggled financially in the early 1970s after the initial outlay of enormous sums of capital to build the system suddenly found it much easier to attract new subscribers. No longer was cable the small underdeveloped, subservient poor sister to the conventional networks. Cable operators found that in addition to the one-time installation fee, usually then about $30, and a fee of about $10 a month for 12-20 channels, subscribers were willing to pay an additional monthly fee to see recently released movies and otherwise unavailable sporting attractions, all without commercials. The effect of satellite development was that cable television moved heavily into urban constructions. The pre-satellite cable penetration of 35 per cent in urban areas would not have warranted most urban constructions, but in response to satellite programs 80 per cent of that 35 per cent took an additional pay service.[28] Home Box Office, owned by Time Inc, emerged with a successful formula for the packaging and distribution of pay television, supplying the same programs to several cable systems, who could then repeat them several times a month. The switch to satellite distribution confirmed HBO's success: it reached one million subscribers to become profitable in 1977 and by 1983 had over 12 million subscribers.[29]

The popularity of pay television in the USA spawned problems for the cable-television industry as a whole. Impressed that they could draw $40 per month from some subscribers, cable-systems operators became overconfident and embarked upon a building spree in the late 1970s. To construct a cable-television system in the USA requires the approval of the municipal council, which grants franchises to its chosen operators. The early 1980s saw a war for new franchises, with major system operators bidding unrealistically for seemingly lucrative urban markets. With rising interest rates, spiralling costs, and consistently delayed construction programs the American cable-television industry overreached itself in the early 1980s. From 1974 to 1984 the cost per subscriber to the operator, to install a typical cable-television system, leapt from $120 to $900.[30] Municipal authorities revelled in this seller's market and they came to expect as a minimum competitive franchise bid a 54-channel interactive system, with extensive access, community and educational facilities.

Programming for cable television has some important differences from that of conventional over-the-air commercial television. Cable systems provide more channels and thus facilitate the fragmentation of traditional mass markets. This 'demassification' of television means change in the broad sense from few programs distributed widely, to many programs distributed narrowly. The new distribution system, called narrowcasting, may mean dozens of specialist channels focusing on serving the television audience's vastly diverse interests, affinities and tastes. With cable television, viewers pay the programming distributors – the cable company – for what they get, as a monthly account or on a program-by-program basis. The nexus of program controllers to advertising is thus broken, a profound structural change. Since with basic cable the audience pays for a range of programming services, the cable company offers as many programs as they can find to put on the system. Instead of a small group of elitist managerial program decision-makers with prime concern that they select programs that will best deliver the audience numbers to advertisers who pay them, their commercial motivation is rather to find as many programs as possible, charged directly to the audience.

Cable television in North America has provided a vehicle for community involvement in television programming. As soon as the cable-television industry was born, cable operators and media-access groups began to provide programs about local affairs for broadcast on an available channel. They used inexpensive video, were unprofessional by conventional production standards, but they brought local issues and events to their subscribers. The program makers and participants were ordinary people interested in sharing information within a particular community. This is a different use of the medium. The history of Australian commercial television shows that this medium is the least participatory medium of all. Few members of the community are involved in programming decisions, fewer have access of any kind to make programs: part of a phenomenon of media as something 'that is done to us'. We have much to learn from others about community television, particularly from the Canadians.

The importance and potential of a community channel were

recognised by the Canadian government when the Canadian Radio-Television Commission, now the Canadian Radio-Television and Telecommunications Commission (CRTC) was established in 1968. Many members of the CRTC were convinced that the community-television concept could be developed by the cable-television industry into a valuable social service to the Canadian people. The place of community channels in a cable system has always been central to their assessment of particular operators as the Canadian Broadcasting Act, which applies to cable as well as over-the-air broadcasting, specifies that an important function of television programming is to contribute to the development of the sense of Canadian identity. No such admirable statement of intent appears in the Australian Broadcasting and Television Act.

The community-channel concept developed slowly at first. Most cablecasters devoted the bulk of their time and money to expanding their systems. It was during the period 1970-75 that the number of community channels increased dramatically, as did the level of public awareness about their potential. The community channel has always been a no-charge communications vehicle funded by the licensee. In 1985 the CRTC proposed that 10 per cent of gross subscriber revenue be put toward financing the community channel, a proposal not strictly enacted, but used as an important guideline to the industry. The track record of companies in this area, their expenditure and desire to encourage rather than tolerate the community-channel concept, is important with regard to CRTC sessions concerning licence renewals, subscriber rate increases and other requests made by the cable companies to the Commission.

The Canadian community channel is made up of two programming components, local origination and access. Local origination is where the cable company telecasts local sports or covers a town-council meeting. Access means that community members either have direct access to video equipment to produce what they want, or they have access to producers who are responsible for the technical side of production, while the public supplies the ideas, content and talent. According to a CRTC survey in the late 1970s, community cable channels alone were staffed by 398 full-time staff and 273 part-time staff: in addition over 3600 volunteers were involved in cable-television systems.[31] Analysts are in general agreement that the American Cable Television industry is currently at the end of a shake-out period, with many small established operations in good financial shape, but with some operators who became embroiled in the recent urban franchise wars in a precariously overreached financial position. The industry has faced major problems in the 1980s – high capital costs and interest rates, substantially increased systems-construction costs, serious marketing problems, and intense competition from the established networks, as well as from new alternative distribution systems. Warner Amex reported a loss of US$47 million on revenues of US$308 million in 1982, and its estimated losses for 1983 and 1984 were $25 million each year.[32] The strategic financial decision for several major corporations with new urban franchises is what level of loss they can continue to endure.

On the question of whether cable television has realised – or will realise – the dream of diversity, the jury is still out. Lower advertising revenues and fewer subscriptions than anticipated have forced the closure of several specialist-program networks, including CBS cable, an arts and leisure channel and Westinghouse/ABC's Satellite News channel. RCA's The Entertainment Channel, with losses estimated at $35 million in its first year, was absorbed by the former ABC/Hearst's ARTS channel, to become the Arts and Entertainment Network.[33] Initially, the Entertainment Channel offered an exclusive BBC package of programs (40 per cent of its schedule), original American series, Broadway adaptations and foreign films. On balance, cable and pay television in North America has essentially provided more channels, more programs, more services and more images. Yet this new communications technology has not spawned new forms of television, or democratised television, and certainly has not banished the established commercial-television networks.

What then of the Australian prospects for cable and pay television? At the very least, the introduction of a CTV system in Australia would bring a greater range of channels and choice than the present limited offerings, a maximum of five channels, and then only for Sydney and Melbourne. A multi-channel, basic-cable subscription-television service combined could offer an Australian city a choice such as:

Channel 2 – ABV 2 (present ABC Channel Two – 'must carry' rule)

Channel 3 – House of Representatives or Senate (direct telecast, as in USA or Canada)
4 – Movies or Australian drama (re-runs, as part of the basic service)
5 – Public Broadcasting Service (documentaries from the USA)
6 – Local Community Information
7 – Network 7 (commercial channels remain in)
8 – Adult Education Courses (public lectures and courses).
9 – Network 9 (commercial channels remain in)
10 – Network 10 (commercial channels remain in)
11 – CNN (USA 24-hour news service)
12 – Community Channel One
13 – Community Channel Two
14 – Multi-Lingual Community Channel
15 – Video Pool (home and education video programs)
16 – Colleges and Universities of the Air (public lectures and formal educational study)
17 – The Sports Channel (Australian and international events)
18 – Videotex
19 – Time, weather, program guide
20 – Home Box Office (US pay-TV network)
21 – BBC (pay TV)
22 – AFC Films (pay TV, from Australian Film Commission)
23 – The Arts and Entertainment Network
24 – Children's Television (from US Nickelodeon and Children's Television Foundation)
25 – The Music Channel
26 – ABC Second Channel
27 – Videocassette library
28 – SBS Channel 0/28 (present multi-cultural television)

What though of the vexed policy issues involved with the introduction of cable and subscription television in Australia? At the request of the Fraser administration, the Australian Broadcasting Tribunal conducted an extensive public investigation of the possible introduction of CTV and RSTV and recommended the introduction of both systems 'as soon as practicable' in their five-volume report in August 1982.[34] This report argued that Australia was on the threshold of a new era in communications, and justification for the decision to recommend to proceed was initially on the grounds of the entertainment benefits of CTV and RSTV, but the real significance of these new technologies lay in their 'strategic value for Australia's long term participation in the information age'.

To be fair-minded about this report, the members of the Australian Broadcasting Tribunal faced the daunting prospect of assessing not only new communications technologies, but fundamental changes to Australian broadcasting without any social objectives having been designed for the system as a whole. As we have seen with community television in Canada it was supported by the regulatory authority because such developments were consistent with the stated ideals for Canadian broadcasting, endorsed by Canadian parliaments. In the Australian context, our regulatory authority had no comparable frame of reference to work with and it essentially adopted a pragmatic policy stance. The ABT report stated that 'the social, technical and economic arguments against the adoption of cable television were outweighed by the favourable arguments.'[35]

It appeared that members of the Tribunal made an early judgement to support the introduction of cable and subscription television, and then put their trust in the demand studies and consultants' reports to substantiate their recommendations.

The ABT hearings indicated more interest from companies than was the case in the lead-up to the introduction of commercial television itself in the 1950s. Twelve organisations expressed interest in owning CTV reticulation systems, including Telecom, major Australian media groups, US and Canadian cable companies, and three large Australian companies who were newcomers to media – Myer, Elders IXL and Amatil. In addition, 72 submissions indicated their interest as potential operators (though some for one channel only), 94 submissions indicated interest as program suppliers and 31 submissions identified an interest in the provision of equipment for the operation of new television services. For RSTV, 31 organisations expressed their interest as a possible systems operator in Australia.

The ABT report embraced the mood of broadcasting deregulation and greater privatisation by recommending that the ownership, control and operation of Australian CTV and RSTV systems be the exclusive province of private-sector participants. It recommended that the ABC and SBS *not* be permitted to own, control or operate CTV or RSTV stations, though it was acknowledged that the ABC in particular would be in a strong position in production, packaging and supply of programs to private-sector-systems operators. The Tribunal raised two fundamental questions about the ABC as a subscription-television operator. First, since a capital outlay of $100 to $150 million would be required for decoders, they asked whether public funds of this magnitude should be committed to an entrepreneurial venture with 'its attendant degree of financial risk.' Second, they raised doubts about the ABC's ability in marketing, pointing out that 'it has only limited marketing activity experience.' This is especially crucial to an operation where overseas experience would indicate that RSTV systems need efficient and skilled marketing techniques to survive, let alone succeed.

On the vexed issue of ownership of the cable-reticulation systems, the Tribunal did not opt to endorse the continuation of Telecom's national common-carrier status. The ABT rejected Telecom's proposal to act as the exclusive common carrier to install and maintain the cable-reticulation systems for the provision of CTV programming. Members of the ABT had agonised over the issue of the extension of Telecom's monopoly to cable systems and noted the fervent opposition from several private-sector prospective operators. Instead, the ABT recommended 'an open and competitive approach whereby no organisation or group is given the exclusive right to own CTV reticulation systems but ownership is permitted on as wide a basis as possible.'[36] They added the firm recommendation 'that Telecom *not* be given the exclusive right to own CTV reticulation systems.'

Curiously, the Davidson inquiry into Telecom, which ran partly simultaneously with the Jones inquiry into CTV, produced a different finding on this central issue.[37] The Davidson committee rejected the prospect in the Jones report for cable-systems ownership to be from the same organisation as the television-program provider and stressed the important principle of the separation of 'system' from 'information content'. Davidson warned that if the principle of separation was not realised there would be a resultant need for complex regulation to ensure fair access for other telecommunications-service providers, and for competitive information services. Moreover, cable-television reticulation systems needed to be integrated into the overall planning of the total national telecommunications structure. Though not stated explicity, the Davidson committee – established to find ways to privatise the Australian telecommunications system – endorsed Telecom's monopoly for cable-television systems in Australia.

On the question of the ground rules for prospective licensees, the ABT did not necessarily preclude existing broadcasters from becoming cable operators. It argued that the licence-granting authority be given discretionary powers based on factors such as prescribed interest levels, total media interests and the viability of existing broadcasting services. Chairman David Jones, in a public address following the release of the report, said that the present incumbents would probably carry 'more lead in their saddle' than would a fresh prospective licence candidate. The ABT did not, however, isolate in this report the problems of concentration of broadcasting ownership as they did later in their SPS satellite report, nor seize upon the prospect of using CTV as a policy mechanism for ownership diversification.

Somewhat surprisingly, on the question of foreign involvement in CTV licences, the ABT found difficulty in deciding what foreign-ownership policy should be applied. They did not simply adopt the existing provisions of the Broadcasting and Television Act relating to foreign ownership or control, essentially an upper limit of 20 per cent foreign interest with a maximum of 15 per cent in one hand, for CTV. They stressed two 'special considerations' in the area of foreign ownership and control. Companies that have been 'naturalised' deserved special consideration, they argued, and relaxation of the present restrictions ought to be considered in the early stages of the development of the cable industry. Such considerations were not considered appropriate for RSTV where, according to their report, we merely ought to adapt the present provisions of the Broadcasting and Television Act.

The proposed licensing system was to be based on competitive bidding, with applicants making promises of performance about their future programming offerings – a system tried and failed for conventional commercial television. The ABT report recommended that licences for CTV operators be granted for as long as

15 years, with a 10-year renewal assessment, and for RSTV operators, a five-year licence with renewal every three years. Advertisements would be permitted on the basic cable channels, but not permitted with RSTV encoded signals. On the vexed programming areas concerning the extent of Australian content and children's television, the report suggested statements of intent by prospective operators as part of their licence application. However, to its great credit, the ABT called for a statutory requirement in an Australian cable system to provide channels for each of community, education and children's programs. Not only could this have been a way that some form of broadcast community television was at last introduced into Australia, but it would also have acted as a stimulus for programming diversity and greater local production.

The Fraser administration stalled on the report, failed to state a policy position publicly, and then got caught up in the 1983 election. Labor's shadow Media Minister, Senator John Button, strong opponent of cable television, which he regarded as an innovation 'very much in the silk department', declared during the 1983 election campaign that 'there will be no cable in the foreseeable future if the Labor Government is elected.'[38] Given the arguments he had presented to the ABT public hearings, this was absolutely predictable. A summary of his plausible case against the introduction of cable television into Australia is worth quoting, though it did show lack of appreciation of the potential and prospects for change that cable could bring to the Australian broadcasting system.

In presenting evidence to the Australian Broadcasting Tribunal's *Inquiry Into Cable and Subscription Television Services and Related Matters* Senator Button offered this justification against cable television:

> Cable will involve, if introduced on a large scale, a very significant commitment of capital resources . . . I myself would doubt if in terms of social and economic priority it is either wise or necessary to appropriate those resources in Australia to cable television . . . Secondly, any demand for cable will of necessity be created demand . . . The third point is that I find most of the arguments in favour of the introduction of cable television largely concerned with what I would call self-interest arguments but they are supported in part by what I would call a slavish adherence to a technological imperative. That is to say because it exists, it

is necessary for us to have it in Australia . . . There is a very strong tendency . . . to extrapolate US experience to vastly different conditions in Australia . . . I am not myself interested in introducing new systems here based on new technology which are going to benefit perhaps one third of the community to the social exclusion of others. The other question that is related to that is the question of what minorities in Australia will specifically benefit from cable television and how big are those minorities . . . The solution to ethnic community problems is not seen as being in segregation which cable would essentially provide if it were to cater for specific ethnic communities . . . I would really summarise by saying that at the present stage of understanding about what cable would do for Australia I find that in social terms, it would have much the same social utility for Australian society as, for example, poker machines.[39]

Cable television has been a dead issue under a Labor Government in Australia. Not so RSTV, however. The first Hawke Government twice baulked at accepting recommendations for the introduction of pay television. The problems emerging for pay-television operators overseas have been noted, and the difficulties of selling the concept of 'user pays' to an Australian electorate which has only ever had 'free' television have been well understood by the pragmatists in Cabinet.

In February 1984, Michael Duffy, Minister for Communications, called for expressions of interest from potential operators and issued a broad set of guidelines for RSTV in Australia. This document stressed that the Government had yet to decide whether to introduce RSTV, but that the Government wanted to consider 'whether RSTV could boost Australian film and television production and also have an impact on the development of Australian technology.' Duffy received 42 expressions of interest from organisations who wished to be involved with different aspects of RSTV, from prospective manufacturers of decoders, to program distributors, to interested system operators.

The Myer Emporium Pty Ltd submitted a detailed proposal to service the Melbourne metropolitan area, intended to be followed by an extension of the service to Tasmania and outlying areas of Victoria, using the south-eastern footprint of the domestic satellite system. Alan Bond's STW Pty Ltd in Perth proposed a 24-hour-a-day movie channel. Television Australia-Satellite Systems Ltd proposed an unusual mix of a dual community-commercial television service to remote areas using four spot-beam transponders of

Aussat, where commercial-television programming would be offered from 4pm to midnight and community programming between 8am and midday. Other organisations who registered their interest included the US operator Oak Industries, Elders IXL, James Hardie and Sons, and the cinema exhibitors Hoyts and Village Theatres. In addition to these prospective participants, the new board of the Australian Broadcasting Corporation had earlier made a submission for a national ABC pay-television service to be licensed. Meanwhile, the commercial-television industry is predictably opposed to the introduction of RSTV in all areas.

There are some tough issues to be resolved on the possible introduction of RSTV in Australia. Is it a service that is needed and by whom? Under what conditions might licensees operate? Has the videocassette-recorder invasion undermined pay-television markets? Would RSTV provide any real programming diversity? Would employment prospects in the video, film and television industries be enhanced, or might it lead to a drop in the local production of films and television? Would it introduce unfortunate inequalities of access to broadcasting services throughout Australia?

11 The Electronic Estate: Politics Not Policies

Technology and communications policy in Australia has been characterised for decades by an appalling neglect of social philosophy. Governments have lacked an integrated frame of reference for decision making in the communications and wider technological fields, based on thorough social investigation, strategic planning, strong public awareness and a careful assessment of options for the future.

Ad hoc decision making is the inevitable consequence of an essentially irrational political process, from which clear comprehensive policy statements fail to emerge as blueprints for action. The response to technological choices by governments in Australian communications policy has essentially been either a series of short-term pragmatic solutions, or the shelving of decisions in complex policy areas. Inevitably policy errors or neglect of the past must be paid for by the community as a whole. There is a tendency in the communications field for major past bad policy decisions to establish their own ratchet effect – discussions which then permit motion in one direction only.

The overall position of technology and communications policy is a chaotic frame of reference, with a generally ill-informed community, and a political process ill-equipped to come to terms with major issues for the future.

A small elite group of policy makers holds the balance of power in the key decision-making process. This group is composed temporarily of a few politicians who are members of the government of the day, together with their key advisors, and the more permanent senior members of the bureaucracy, with selective inputs from commercial interests and from effective pressure groups. A prime paradox of public policy is that major policy decisions are made in forums which are essentially private and closed, and within institutions able to wrap themselves in secrecy. In the fields of technology and communications, there is no permanently established mechanism whereby a wider cross section of the public can effectively offer an input into the policy-decision-making process and later effectively analyse the outcomes. Moreover, the mass media are structured in such a way that relatively few people have access, and the accepted conventions of presentation restrict discussion and exploration of complex subjects. In the context of debate about technological change and the future, the political system as a whole has the effect of inhibiting curiosity and limiting public awareness.

The notion of an information or post-industrial society suggests that a new kind of society has emerged. If we choose to name it from the nature of its ruling class, then it ought to be called a technocratic society in which technocracy 'signifies a social order organised on principles established by technical experts'.[1] For Habermas, we now live in an era of technocratically managed capitalism. Politics has, in his view, become basically reduced to a question of who can run the economy best – perceived as merely a matter of technical decision making. Modern capitalism has seen the emergence of pragmatic, technocratic government, where the mass of the population have become alienated from a political system which has failed to maintain its own narrow self-defined objective of sustained economic growth.

Technocrats have become both policy advisors and policy makers. They have had a dominant influence on the plethora of public inquiries and major government reports in the departments of science and technology and communications. At face value there appears to have been an exhaustive process of public consultation in these fields in Australia over the past decade. We have witnessed the broadly based Myers Report of the Committee of Inquiry into Technological Change (1979), inquiries into telecommunications institutions from Vernon (1975) and Davidson (1982), multiple satellite reports and inquiries such as from Bond (1977), the White Task Force (1978), the Working Party Report (1979) and the Australian Broadcasting Tribunal's SPS Report (1984), inquiries into new communications technologies such as the Jones ABT report on cable and radiated subscription television services (1982), the ASTEC report on videotex (1983) and reports on information technology from Scott-Little management consultants (1984) and from the IAC (1984).

These inquiries and reports have, of course, provided a valuable body of information. However, the manner in which many of these investigative and advisory organs have been constituted has severely limited their ultimate value to the policy-decision-making process. Too many of them have been conducted without any adequate social frame of reference. They have concentrated on the introduction of a new technology (cable, pay, satellites), or individual institutional studies (ABC, SBS, Telecom) or industry-sector reports (the computer industry or information technology). Even in technical terms alone, these inquiries have offered a piecemeal approach of compartmentalisation at a time when the technological tendency is towards convergence. In some cases there has been a lack of co-ordination between inquiries which have run simultaneously, and some inquiries have presented contradictory findings and recommendations on major policy issues. For example, on the central issue of Telecom's role in the reticulation for cable-television systems, Jones and Davidson filed dramatically opposed recommendations on the proposed role for Australia's national common carrier.

Most inquiries were given limited resources, especially for research, and most members worked on a part-time basis towards a tight deadline. Those inquiries which failed to deliver a handy check list of straightforward recommendations tended to be attacked most of all. For example, the Jones 1984 satellite inquiry[2] – the first time the Australian Broadcasting Tribunal had been formally consulted about the implications for commercial broadcasting of a domestic satellite system – suggested that the issues raised required further parliamentary debate and the development of long-term coherent broadcasting policy. What is expected though from government inquiries is action and decisiveness from technical experts, even if the advice is irrational or wrong.

The multiplicity of government reports in the fields of science, technology, media and communications have collectively failed to locate Australia's technological position in the broad context of its contemporary level of economic development. Australia's technological industrial base, which has progressively contracted during the last thirty years, is now sometimes referred to as an industrial museum. While information services are booming, the manufacturing base of the computer and telecommunications industries –

the electronics-manufacturing industry – is a decimated indigenous business. We have become progressively more dependent on higher levels of imported electronics manufactured goods. In Australia, local manufacturing industry does not form an integrated high-technology infra-structure. Australia has an extremely low level of investment in industrial research and development, and inadequate technological transfer between its research laboratories and industry. We have chosen to ignore long-term strategic economic and industrial planning, in both the public and private sectors, and to remain as a technological client state.

Australia needs to develop a range of industrial policies which are fundamentally nationalistic. Industries associated with the business of media, information, communications and telecommunications collectively now form a sector of substantial economic growth. Those who generally argue that the information economy is the engine for future growth throughout the world have been encouraged by estimates that OECD countries will devote $800 billion over the next decade to the development of new communications systems.[3] Investments already in place in Australia would suggest that the 'information boom' is already of greater magnitude than what the anticipated 'resources boom' was expected to be.

The notion of an Australian information economy is not merely a prediction, but an economic process already well underway. The benefits of this development, however, will not be fully realised by the Australian community if the present structural trends in the economy continue. As we have seen, in the information technology field, Australia has been colonised, while the products are essentially those of trans-national corporations who do not invest in R & D in Australia and who affect our appalling balance of payments of high-technology goods. We have high imports of the products of trans-national corporations, limited high-technology export products of our own, and the people employed in Australian subsidiary companies remain for as long as it suits the goals of the international parent corporation.

An Australian MITI

What is necessary to redress this economic imperialism is a directed economic model, with substantial public investment, and a series of government preferential industrial policies directed towards designated growth areas for Australian owned and controlled companies. The fashioning of future economic strategies for Australia's electronic estate must be assessed in terms of the continuum alluded to earlier – from the freewheeling, laissez-faire, marketplace dominance on the one hand, as opposed to a state-directed, strategically planned economic model on the other, with its prime focus on support for indigenous-based industrial ownership and control. The 1960s and 1970s orthodox Australian economic philosophy, put simply, that government need only frame broad economic policies designed to engender a climate of confidence, and thereby stimulate private-sector growth, failed to establish a stable national industrial infra-structure for the future. An essential laissez-faire economic model for Australia ultimately produced massive de-industrialisation and associated record unemployment.

Australia needs a strategic economic-policy authority similar to Japan's Ministry of International Trade and Industry (MITI). As we have seen earlier, Australia has a remarkable record of scientific invention, yet a deplorable failure to convert scientific innovation into its own successful commercial ventures. The Japanese industrial position has been essentially the opposite – rather unimaginative innovators but brilliant industrial entrepreneurs. The Japanese strategy has been essentially towards targeted concentration of industrial development, where the MITI has provided the state-directed planning and finance necessary to develop a secure indigenous industrial base. It must be stressed that, though government is the prime architect of strategic economic planning on this model, it is the private sector that is supported, stimulated and rewarded for its successful industrial ventures. The essential MITI strategy is that by promoting a few corporations to monopolise the information industry, Japan can be assured that these firms will be successful on a global scale. Though a great deal more thinking needs to be done on the concept, an Australian version of MITI may be the only way in which we can overcome consistent failure to develop our missing stage of indigenous engineering and commercialisation, a necessary pre-requisite for a stronger Australian export-based information economy.

In the context of an information economy, an Australian MITI needs careful consideration of how best the present ministries and departments of the bureaucracy could be brought together in a 'super' economic and social development ministry. At least three ministries are vital to such a process of integration - Industry and Commerce, Science and Technology, and Communications, though elements of the portfolios of Trade, Employment and Industrial Relations, Education and Defence Support would also have to be involved. Actually, such a process would be entirely consistent with the ALP platform, which emphasises the inadequacy of total reliance on market forces, the problems of foreign investment, and which declares support for emerging industries in such a way that technological innovation and development is based on an economically nationalistic approach.[4]

There are several major problems associated with gaining acceptance of the concept of an Australian MITI. Commitment at the highest levels of the elected government is not in keeping with the pragmatic, short-term considerations which have come to characterise contemporary Australian politics. An overall publicly planned economic model would be interpreted in some quarters as an extension of regressive tariff-protection policies for particular industries. Such a proposal would run into entrenched opposition from within the federal bureaucracy. Support for a free-wheeling economic model has manifested itself from the level of the Industries Assistance Commission - which has argued against government support for the indigenous Australian computer and software industry - to the more influential higher levels of Treasury, which have blocked bold investment initiatives such as those proposed by Barry Jones. Bureaucratic conservatism against economic vision and long-term strategic planning has been encapsulated in their slogan of objection that 'governments should not be picking winners'. The point about the contemporary Australian economy is that information industries *are* winners, but not *our* winners. The need for a publicly directed economic model is to ensure that we plan now to guarantee a secure indigenous industrial foundation for the future.

A case exists for the retention of an essentially public model in the Australian telecommunications industry. In the drive towards privatisation in telecommunications, the significance of long-term public investment appears to have been forgotten. It is the nature of new services in telecommunications that substantial public investment is accompanied by a long lead-time before the rewards are realised. Three overseas historical examples support the point. America's world leadership in communication satellites in the mid 1980s, by corporations today such as Hughes Communications International, was derived from the massive public funds injected into the American space program during the 1960s. Videotex, invented at the research laboratories of British Telecom and launched as Prestel, was a multi-million-pound investment by British Telecom, who was prepared to invest over a long time scale for eventual returns. If videotex becomes a major new technology in the communications field, then the rest of the world will have benefited from the British public investment. France now has the most developed data-transmission service in the world – the result of the substantial public investment by the French PTT, and guaranteed by the monopoly of their domestic market.

Public-telecommunications authorities around the world are often the single largest investors in their domestic economies, and Telecom Australia is no exception to this general rule. Large-scale, long-term investment is necessary to plan for an efficient telecommunications national network, and this appears to be only possible with public organisations or utilities. Major telecommunications developments, such as large switching exchanges or fibre-optics networks, need careful long-term planning to ensure that an acceptable return is made on investment. It will always be possible for a competitor to offer an alternative limited service, such as a microwave link, which would be cheaper than the PTT price for a comparable service. However, the question of aggressive competition is more than merely siphoning away revenues from Telecom, but inherently carries the risk of the primary carrier adopting a cautious approach to long-term network investments. It will not be possible for Telecom to expand its present terrestrial network at affordable costs to the consumer and be in the forefront of latest technical innovations unless it is guaranteed an absolutely secure financial base, with assured growth prospects.

Most developed countries have linked the monopoly position of the PTT with strong national government procurement policies to support local industry. Most European PTTs have strong manufacturing support as a condition of their purchasing contract, a situation which has not changed with the free-trade EEC. Wholesale deregulation must work against a policy of national procurement. In Australia, with an exceptionally high degree of foreign ownership of its manufactured goods, extensive deregulation would be even more disastrous for local telecommunications-equipment-manufacturing companies. What Australia needs in its information-based industries is large-scale procurement from indigenous companies as part of a national strategy towards long-term economic benefit. We need to work towards large-scale government purchase of local computer equipment, Aussat purchase of locally manufactured satellite hardware, and substantial OTC and Telecom purchase of telecommunications equipment from Australian owned companies. Telecom, in particular, with its outlays in excess of $1000 million annually on telecommunications equipment, has a central role to play in this context. Telecom must expand its indigenous procurement policies, be encouraged to establish joint commercial ventures with Australian owned engineering companies, and embark on a pump-priming role for the Australian telecommunications-manufacturing industry. Telecom must work in close consultation with the private sector, adjust to specific new opportunities for private companies, but essentially maintain its central strategic position in Australian telcommunications-industry developments.

The approach sketched above for the Australian telecommunications industry ought to come under the policy umbrella of a MITI, though no substantial changes to the charter of Telecom, Aussat and OTC is envisaged. The MITI would need a top-level policy-advisory vehicle not attached to the narrow confines of one ministry or one government department. The process of policy formulation in the computer-communications field needs to be urgently re-evaluated. We can no longer rely on a plethora of inquiries into isolated components of technological change. How can we bring together disparate investigative and research activities in forms that are intelligible and valuable to the community? How can we work towards a more rational approach towards investigation of the processes of technological change in the context of Australia's future?

The notion of a Commission for the Future, urged by Barry Jones, is a promising venture. A role of the commission is to draw up scenarios for the next two or three decades into the future, and to put forward options from which ideally the community can make its own informed choice about future directions.

The Commission for the Future was envisaged as a body to

contribute towards raising public consciousness of technological issues rather than as an advisory body for government. Yet an Australian MITI would need a top-level advisory body on social policy. A Commission for the Future, if given a broader charter, would be able to pull together a great deal of applied social investigation and explore more thoroughly the relationships between technological change and contemporary Australian society.

If such a body could produce an Australian version of the French government's 1970s commissioned blueprint for social change, Simon Nora's and Alain Minc's *The Computerisation of Society*, an immensely valuable advisory function would be served.[5] This would not only avoid the problem of the too narrowly focused public inquiries of the past, but such a commission would be able to cut across many government departments to suggest future strategies without parochial attachment. There are precedents for the success of such an approach. Justice Michael Kirby's Australian Law Reform Commission achieved rare insights and valuable blueprints for change across a mosaic of complex legal social issues. Why not an Australian Commission for the Future as a broadly based technology advisory committee which is accepted by government as having a central role in top-level polic formation?

Technology can never be neutral – it has no independent existence. Technological development occurs within an ideological framework and through economic and social institutions. Is there not just a chance that we can begin to fashion our technological future?

Open Media

No broadcasting-policy objectives for Australia are set out in the *Broadcasting and Television Act*. How, really, can responsible decisions be made and implemented about new media in Australia in the absence of national strategic objectives? First, we need a media and broadcasting system which is not only effectively owned and controlled by Australians, but also where that control is not concentrated in too few hands. But past errors of legislative judgement pose monumental problems for change in the area of concentration of ownership. A second closely related objective is the need for a media and broadcasting system with a range of controllers of different kinds, or one which offers an 'ideological mix' of key participants. That is, the system should offer a range of controllers who have different perceptions of the world and whose perceptions are reflected in their cultural products. With commercial broadcasters, public-service broadcasters (the ABC and SBS) and public or community-radio broadcasters, an embryonic systems flexibility exists, though public television needs to be added as a new sector. Each sector needs to be assured of a financially stable place within the system, and changes are needed to redistribute income within the various sectors of Australian broadcasting. Third, the system must provide increased opportunities for employment within Australia, and for greater access and community participation in media. A more democratic media system would stimulate the creative opportunities for participants through its openness and accessibility. Fourth, greater geographical equity in the distribution of cultural products needs to be achieved, with program services made much more widely available throughout Australia. Fifth, the communication needs of audiences needs to be more investigated, and the new technologies utilised to satisfy heterogeneous audiences needs. We have had a deplorable pre-occupation with means and techniques, based on hardware considerations, rather than beginning with consideration of the human condition.

What Australian broadcasters have to offer their public in future depends largely on industry power structures. We have seen that Australia has the most concentrated pattern of media ownership and control in the democratic world. Moreover, there are some indications that the development of new communications technologies, and particularly the introduction of an Australian domestic-satellite system, could lead to even further concentration of media ownership and control. The Aussat system has the potential to convert present commercial-television markets into virtually one Australian commercial-television-industry system, dominated by three Sydney-Melbourne based networks. Instead of making decisions about the future of Australian broadcasting based on strategic policy making, the government has chosen instead to offer some broad policy guidelines in response to problems. Mindful of Aussat's commercial responsibilities, and sensitive to the political implications if the satellite project did not become financially viable, the government has found itself unfortunately dependent on the corporate financial strength of the three commercial television networks. The inherent expense of the domestic-satellite project has meant that what has been ignored is the notion that Aussat could cross-subsidise users in order to get greater diversity into the broadcasting system. A HACBSS 2 hybrid model of small users – including community, educational and ethnic broadcasters – never emerged because of resistance from commercial networks and because those groups lacked adequate assured income. The established principle for Telecom's terrestrial network – that profitable services cross-subsidise other needs – has barely been considered in terms of the introduction of new satellite technology for Australia.

Once the commercial-television networks have become major Aussat customers they are likely to entrench their strategic position. If Australia's public debt continues to escalate, and if a change of government in the future led to stringent monetarist policies, a public-service broadcasting organisation faced with substantial budget cuts, such as the ABC, might no longer be able to fund the costs of transponder hiring. Who then, amongst the few financial eligibles, could take over the HACBSS 1 with its four expensive high-powered transponders? This is speculative, but an extended role for commercial-television networks is possible in a resurgent era of the privatisation of broadcasting.

The networks proposal to the Australian Broadcasting Tribunal's SPS Inquiry would have placed them in an ideal position for the eventual introduction of direct-broadcasting satellite services. Although the Australian satellite system was developed as a fixed satellite service rather than as a broadcasting satellite service, developments in satellite and receiver technology led to consideration at that inquiry as to whether the major television stations should be allocated the second set of high-powered transponders for them to transmit programs to cover the entire continent. A design modification in 1982 would have allowed the 30-watt transponders to be switched into national beams for seven out of the eight high-powered transponders on the first two satellites. This could have allowed remote-area dwellers to decode a commercial television signal, and the evidence submitted to the ABT suggested that the cost of direct homestead commercial-television reception would be in the range of $3500 – $5000.[6] As explained earlier, this network proposal was rejected by the ABT and government in favour of regional consortia with access to the 30-watt spot beams. It seems then that the networks will have access to high-powered transponders only when the third satellite later becomes operational, provided no spot beam is left unallocated by the failure of a regional-television consortium to take up its option. The next generation of satellites in Australia will probably possess a direct-broadcast capacity. Programs will be able to be beamed to roof top dishes from one distribution system. Some futurists envisage a stage after Australian DBS when signals will be delivered internationally via satellite, and where a range of signals will be able to be received throughout Australia from many different nations. There was evidence at the Australian Broadcasting Tribunal's satellite inquiry that the networks wanted access to Aussat's high-powered transponders to absolutely secure their position in any possible future DBS Australian development.[7] In the short term this appears to be a strategy based on 'protection of the flank' by commercial-television operators, but given the rapid rate of technological development in broadcasting in recent years, DBS may become the entrepreneurial frontier of Australian commercial television for the 1990s.

If the problems associated with the excessive dominance of a few networks-controllers are to be faced in the immediate future, what structural options for change are possible? In theory, the only way of fundamentally coming to terms with the problem of excessively high levels of concentrated ownership and control is to repeal the present *Broadcasting and Television Act*, divest all present licensees of their most valuable asset, a licence to broadcast, and begin a new system with diverse and new participants. This is hardly realistic, given that any government in power would be reluctant to embark on major reform of a communications system upon which they were so dependent to gain office and retain power. Changes to the power structures of ownership and control of Australia's media are unlikely to be achieved, so only modest compromises remain.

Recognising the difficulty of changing current market structures by direct legislative intervention, the Australian Broadcasting Tribunal has floated the concept of federal networking consortia for investigation.[8] The ABT argued that current ownership rules and market structures grew up in another era and are now inflexible. The 'two station rule', intended to minimise concentration of control, has been undermined by the commercial practice of networking among several differently owned channels. However, argued the ABT, if ownership rules were equated to population served, rather than the number of stations owned, groups outside of the Sydney and Melbourne based oligopolists would have the opportunity to substantially expand their markets. This would enable gradual rationalisation of the ownership and control of existing regional markets. The scheme would carry with it the sting in the tail that to be a member of the networking consortium, members would only own and control one television channel, and hence each of the Seven, Nine and Ten networks would have to divest themselves of a station. Such a scheme primarily proposes an economic solution to what is an ideological and cultural problem.

Several changes are possible within the present media and broadcasting structures which could benefit the Australian community. The last decade of Australian broadcasting has seen a widening chasm between resources in the two major sectors: commercial broadcasters have prospered during a period of unprecedented advertising growth, while public broadcasters have faced a quiet crisis associated with reduced funding and lack of adequate resources. As Table 11.1 shows, the financial position of the ABC has substantially weakened relative to commercial broadcasters. In 1974-75 commercial-television revenue was $132 million compared to the ABC's for radio and television of $114 million – a gap of only $18 million between the two sectors. Yet by 1982-83 the respective figures were $816 million (commercial television) and $232 million (ABC) – a gap of $582 million. During this period, with an inflation growth rate of 122 per cent, commercial television substantially outstripped the revenue growth of the ABC. (The problem of using aggregate revenue data as a basis for comparison in the absence of other data is again acknowledged.) During this period the newly created public-radio sector, with aggregate revenues of a mere $4 million by 1982-83, remained resource scarce and poverty stricken. This data shows the general trend in the last decade towards widening inequalities of income between the respective broadcasting sectors.

Herein lies a fundamental problem of how public broadcasters can attempt to 'keep up' with commercial broadcasters in the purchase of high-budget productions and expensive overseas

TABLE 11:1 AUSTRALIAN BROADCASTING: REVENUE $m

	1974-75	1982-83
Commercial Television	132	816
Commercial Radio	57	221
ABC Radio and Television	114	232
SBS Services	–	34
Public Radio	–	4
Public Television	–	–
	303	1307

SOURCES: Annual Reports of Australian Broadcasting Control Board, Australian Broadcasting Tribunal, Australian Broadcasting Commission, Special Broadcasting Services.

NOTE: Commercial Broadcasting and ABC data includes ancillary revenues.

programs. How can public-service broadcasters, such as the ABC, possibly compete with the Packer organisation for international cricket rights, or Murdoch for the Olympic Games, or against any commercial network who wanted to purchase a major drama series such as *The Jewel in the Crown?* A public-broadcasting authority such as the ABC is inevitably in the hands of the government of the day through the mechanism of funding, and a period of contrived austerity, such as during the Fraser administration, can create a series of covert debilitating pressures on the institution. Though crude monetarism may be temporarily shelved in the mid 1980s with the election of a federal Labor government, high public debt will guarantee that public-sector institutions will continue to face competing demands for diminishing public resources.

Hence, in the context of debate about technological change and the introduction of new broadcasting systems, the public-sector broadcasting institutions find that new technological developments in broadcasting are coincident with an era where they are in a financially vulnerable position. As the Australian media moves into a period of change with the introduction of a domestic satellite, the opening up of new regional-television markets, as well as canvassing the possible introduction of cable and pay television, and with direct broadcasting to roof-top dishes a later possibility, industry power balances in Australian broadcasting appear strongly to favour the commercial-broadcasting sector.

Apart from actually changing the present range of licensees, several decisions could be taken that would contribute towards some re-alignment of industry power structures within Australian broadcasting, and provide new opportunities to get new players, of different kinds, into the system. We ought to also use the opportunity associated with the introduction of new communications technologies to provide greater opportunities for more people to become involved with Australian media organisations. The contemporary crises of high unemployment and alienation can no longer wait for economic recovery. In any case, economic growth is likely to be moderate and temporary at best, and it is unlikely to be accompanied ever again by substantial job growth. If we wish to come to terms with how people can find a satisfying place in the community, we must look at a range of policies which challenge the present inequitable distributions of income and work.

The monopoly held by commercial broadcasting over advertising revenue ought to be dismantled in an exercise of some income transfer away from commercial-media organisations and towards public-sector broadcasting. Corporate sponsorship ought to become accepted policy for the ABC, SBS and community broadcasters. This is not to be confused with conventional paid spot advertising which could pose a threat to the editorial integrity of such organisations, and which is clearly intensely disliked by their listeners and viewers. Corporate sponsorship would involve underwriting, co-production and co-financing, where the sponsors would probably only receive a broadcast credit at the start and finish of the particular programs which they supported. In many ways such a policy would legitimise some existing financial practices, such as ABC co-production and co-financing arrangements, and telecasts which have featured advertising hoardings around sports grounds. The potential income must be substantial, especially in view of the ABC's announced intention to seek a 20 per cent share of television viewers between 7pm and 9pm. For some community-radio stations, sponsorship may be their only hope of survival. Corporate sponsorship is not necessarily product oriented and it does not mean that programs must be interrupted frequently by conventional advertisements.

If new media systems are about to emerge in the future in response to new technological options, we need to remember much about the lessons of the past. The dynamism of the private sector in print and broadcasting has undoubtedly offered the Australian community an array of widely accepted cultural products. Yet the public mechanisms of checks and balances, and support for private ventures, have often gone unnoticed. The Australian film industry only flourished in the 1970s after Federal and State government sponsorship. The development of indigenous Australian television drama arose in response to the imposition of local-quotas, and after initial strong resistance from commercial-television management. A local production industry for television commercials only exists because regulation demands that commercials be made by Australians. Children's television has benefited from the introduction of a 'C' classification for approved programs. There is a danger that these kinds of achievements will be swept away by moves towards broadcasting deregulation.

The market forces of Australia's information economy point towards continued concentration, integration and commoditisation. Unless income can be channelled towards new and different media participants, the programming is unlikely to offer anything that is new or challenging. If ever Australia eventually introduced cable-television systems, the prime justification could only be that community needs previously unfulfilled would be served, and that cable television would bring with it a rich diversity of programming. Yet wider scope in programming is likely only to be facilitated through regulatory mechanism. In this context, the Australian Broadcasting Tribunal's recommendations that each cable-television-system operator would have to offer at least three mandatory channels amongst its dozen or so offerings – a community channel, a children's channel and an educational channel – is laudable. Similarly, the introduction of Radiated Subscription Television Services in Australia ought to be accompanied by minimum Australian content guidelines. Regulation in broadcasting must remain as a mechanism to implement the maximum participation by Australian program makers in the industry.

The Australian broadcasting system needs many more participants, in many senses, in both the private and public sectors. In commercial television, for instance, the long period of economic protectionism for regional-television monopolies will at last be broken as a result of the introduction of Australia's domestic satellite system. If this change is associated with greater local community input than has been the case in the past, then much will have been achieved for Australian regional television.

Resistance to changes across media industries as proposed will come from many quarters, and especially from within existing media management. Yet for how much longer can we continue to ignore the toll associated with people who feel estranged from the community of which they could be a part? We urgently need a range of creative policies, in many fields, which seriously examine propositions for the re-distribution of income and work, and which constructively promote community participation. Media are inherently human-intensive - program creation is the principal function of broadcasting - and if the mystique of the media was dismantled and access was encouraged, ordinary mortals could find themselves engaged in the media in satisfying ways. Given the social problems we now face, the 1960s debate about community access and participation in the media is much more relevant to the 1980s.

The computer, telecommunications and media industries are shaping new patterns of production and consumption of information. Paradoxically, the age of technological abundance is also an age of chronic uncertainty. A new kind of industrial revolution is fashioning all levels of social formation. The inadequacy of communications policy, and the associated dilemmas of political decision making, are but a microcosm of the broader social problems of the coming of the electronic estate.

Endnotes

CHAPTER ONE

1. See Richard Butler, Secretary General, International Telecommunications Union, *Communication News*, February 1984 and G. El-Oteffic, 'Planning of Communication' in D. Fisher & L. Harms (eds) in *The Right to Communicate*, International Institute of Communication, Boodle Press, Dublin, 1983, p 160
2. F. Malchup, *Production and Distribution of Knowledge*, Princeton University Press, Princeton, New Jersey, 1962.
3. Daniel Bell, 'The Coming of Post Industrial Society', in *Dialogue*, 11, 2, 1978, p 4
4. Daniel Bell, 'The Social Framework of the Information Society', in (eds), M. Dertouzos and J. Moses, *The Computer Age: A Twenty Year View*, MIT Press, Massachusetts, 1981
5. Alvin Toffler, *The Third Wave*, Kroft, New York, 1980
6. James Martin, *The Wired Society*, Prentice Hall, 1978, p 6
7. Peter Large, *The Micro Revolution*, Fontana, 1980, p 9 & 11
8. Christopher Evans, *The Mighty Micro*, Coronet, 1979, p 94
9. Tom Stonier, *The Wealth of Information*, Thames Methuen, 1983, p 214
10. Available in English translation as S. Nora and A. Minc, *The Computerisation of Society*, MIT Press, Cambridge Massachusetts
11. See annual directory of top 500 American companies in *Fortune*
12. *Fortune*, 24.8.81, p 85
13. American Federation of Information Processing Societies, *Information Processing in the United States: A Quantitative Summary*, Washington, 1977, p 4
14. Wilson, P. Dizard, *The Coming Information Age*, Longmans, New York, 1982
15. *Telecommunications: Pressures and Policies for Change*, OECD, 1983, p 130. The three American corporations are Western Electric, ITT and GTE. Others are (in order) Siemens (Germany), L.M. Ericcson (Sweden), Northern Telecom

(Canada), N.E.C. (Japan), Philips (Netherlands), GCE (France), Thomson Brandt (France), GEC (UK), Plessey (UK) and Italtel (Italy)

16. *Infobrief Research and Technology*, No 233, 5 June, 1983.

17. In the United States District court for the District of Columbia. Civil Action No 74-1698, 1 Nov, 1978, USA (Plaintiff) *v* AT & T Western Electric Co. Inc., Bell Telephone Lab. Inc. (Defendants)

18. *ibid.*

19. *Fortune*, 2 May 1983, p 270

20. Herbert I. Schiller, *Who Knows: Information in the Age of the Fortune 500*, Ablex, New Jersey, 1981

21. Armand Mattelart, *Multinational Corporations and the Control of Culture*, translated from the French by Michael Chanan, Harvester Press, Sussex, 1979

22. Brian Robins, 'MITI Urges More Home Expansion' in *Australian Financial Review*, 20.6.84

23. J. Botkin, D. Dimanesul, R. Stata, *Global Stakes*, Ballinger Publishing Company, Massuchusetts, 1982, p 12

CHAPTER TWO

1. See R. Buckeridge 'Will Information Technology Make Us or Break Us?', Paper delivered to opening of Information Technology Month, Melbourne, 1.11.84. Sources cited include Budget Paper no 10 and Syntec

2. Australian Telecommunication Commission, *Services and Business Outlook for 1984-5*, August 1984, p 6

3. *Information Technology In Australia: Capabilities and Opportunities:* A Report to the Department of Science and Technology by W.D. Scott & Co Ltd in association with Arthur D. Little Inc, July 1984

4. Report of the Victorian Computer Industry Advisory Group, Ministry for Economic Development, Victorian Government, October 1981, p 9

5. Quoted by D. Moore, 'I.B.M. is Still Way Ahead of Industry Rivals', *Australian*, 28.6.83

6. Telecom Australia, *Annual Report*, 1982-83, p 10

7. OTC *Annual Report*, year ended 31 March, 1982, p 7

8. The Law Reform Commission Report, *Privacy*, Report No 22, vol 1, p 45 and the Industries Assistance Commission, *Report on Computer Hardware and Software; Typewriters, Calculating and other Office Machines; Prints and Accessories; Recording Media*, AGPS, 1984, p 58

9. *ibid.*

10. Barry Jones, *Sleepers Wake: Technology And The Future of Work*, OUP, Melbourne, p 242

11. Quoted in a paper presented by Senator Bruce Childs, to the AANA Seminar, Melbourne, 2.3.84
12. IAC *Report on Computer Hardware and Software*, *op cit*
13. B. Jones, 'Technology And Australia's Future' in *National Technology Conference*, Canberra, AGPS p 12
14. *ibid*
15. Philip McIntosh, 'Technology Spawns Its Own Disease', 21.11.84
16. Gerard Mooran, 'Telecom Operators Hit By Strain Injury', Report, *Australian Financial Review*, 25.10.84
17. Quoted by Mr Tickner, ALP in House of Representatives, *Hansard*, 23.8.84 p 256
18. This list was essentially suggested in Mr Tickner's address
19. See two essays, 'Technical Progress and the Social Life-World' and 'Technology and Science as Ideology' in Jurgen Habermas, *Towards a Rational Society: Student Protest, Science and Politics*, translated by Jeremy J. Shapiro, Heinemann, London, 1971
20. *Technological Change in Australia*: Report, 4 volumes, AGPS, Canberra, 1980
21. *Commonwealth Government Task Force on the National Communications Satellite System: Report*, AGPS, Canberra, 1978, *Report of the Committee of Inquiry into Telecommunications Services in Australia*, volumes 1-3, AGPS, Canberra, 1982, Australian Broadcasting Tribunal, *Cable and Subscription Television Services for Australia*, AGPS, August 1982, *The ABC in Review: National Broadcasting in the 1980s*, AGPS, Canberra 1981 and the *Committee of Inquiry into the Monopoly Position of the Australian Postal Commission: Report*, AGPS, Canberra, August, 1982.
22. See Henry Mayer, 'The Cable Decision: Underlying Issues' in M. Armstrong (ed), *New Media: Law and Policy*, University of NSW, 1981
23. See *Sixty Minutes*, 22.4.1984

CHAPTER THREE

1. Argued by Mr Ken Barker, British Minister of State For Industry and Information Technology in *The Australian*, 14.2.84
2. Roger Wittenhall, *CAB* vol 60, no 6, Nov 1983, p 19
3. John Elliot, 'The Need for Deregulation in Australia', The James Kirby Paper 1983, p 13
4. *ibid*, p 13-14
5. Address by the Hon J.S. Dawkins, MP, Minister for Finance and Minister Assisting the Prime Minister in Public Service Matters, to the National Press Club of Australia, Canberra, 28.3.84
6. *ibid*.

7. *Report of the Committee of Inquiry into Telecommunications Services in Australia*, (the Davidson Report) AGPS, Canberra, 1982, vol 1, p 1

8. Robert B. Reich, 'High Technology Industry Policy' – comparing the US with other advanced nations, *Journal of Japanese Trade and Industry*, no 4, 1983

9. *Broadcasting*, 4.4.83

10. *US News and World Report*, 4.10.83

11. FCC Report, no 15650, 7.4.80

12. 'Free the Broadcasting 10000', Address by Mark S. Fowler, chairman, Federal Communications Commission, before the North Carolina Association of Broadcasters, Raleigh, North Carolina, 25.10.82

13. Neilson data in *Broadcasting*, 12.12.83

14. *Broadcasting*, 16.1.84, p 48

15. S. Nora and A. Minc, *The Computerisation of Society*, MIT Press (translated), Massachusetts, 1981 p 6

16. *ibid.*, p 70

17. *ibid.*, p 70-1

18. *ibid.*, p 78

19. *Business Week*, 18.10.82, p 50

20. *ibid.*

21. W. Kirsch, *Transborder Data Flows: A Current Review*, International Institute of Communications, August 1983, p 49

22. 'Videotext and the Electronic Telephone Directory', *Telecom France*, No 5, October 1983

CHAPTER FOUR

1. J. Ford, 'We're Also-Rans in High Tech Export Race, *Australian*, 15.6.84

2. The figure of $3,000 million is used by K.G. McCracken, 'Whither Australian Physics' in *Search*, Vol 14, No 7-8, August/September 1983, p 203: the 90 per cent is estimated in Industries Assistance Commission, *Draft Report on Computer Hardware and Software; Typewriters, Calculating and Other Office Machines; Parts and Accessories; Recording Media*, AGPS, October 1983

3. *Report of the Committee of Inquiry into Telecommunications Services in Australia*, (The Davidson Report) vol 2, AGPS, Canberra, 1982, p 138

4. *ibid.* vol 1, p 122

5. 'The Gwaid Report', quoted in *Australian Electronics Engineering*, February 1984, p 9

6. ATEA Submission to the Davidson Enquiry, February 1982, p 120

7. K.C. McCracken, *op. cit.*, p 202

8. K.C. McCracken, personal correspondence, 1.5.84

9. Ann Moyal, *Clear Across Australia*, Nelson, Melbourne 1984, p 364-366

10. Industries Assistance Commission Report, *Computer Hardware and Software: Typewriters, Calculating and Other Office Machines; Parts and Accessories; Recording Media*, AGPS Canberra, February 1984, p 58
11. *ibid.*, p 66
12. *ibid.*, p 65
13. *ibid.*, p 61
14. Peter Dwyer, 'Hartley Blasts Government Action', *Australian*, 31.8.84
15. IAC Report, *op. cit.*, p 97-112
16. Also, see Greg Earl, 'Shot in the Arm for Domestic Industry'
17. *Australian Financial Review*, 6.7.84. A Report to the Department by W.D. Scott & Co Pty Ltd, in association with Arthur D. Little Inc, *Information Technology in Australia*, vol 1 Department of Science and Technology, July 1984
18. The Davidson Report, *op. cit.*, vol 2, p 130
19. Quoted in the ATEA submission to the Davidson Enquiry, p 123
20. The Davidson Report, *op. cit.*, vol 1, p 125
21. *ibid.*, vol 2, p 146
22. McCracken, *op. cit.*, p 203
23. G.W. Gosewinckel, 'Business Opportunities of the Australian Satellite', address, Regent Hotel, 10.5.84
24. *ibid.*
25. *National Technology Conference*, Proceedings and Report of the Conference held in Canberra, 26-28 September 1983, Department of Science and Technology, AGPS, Canberra 1983, p 1-3
26. *National Technology Strategy, Discussion Draft*, National Technology Conference, Department of Science and Technology, August 1984
27. P. Beard, 'OECD Raps Weak Australian Research and Development', *Australian*, 26.4.84
28. R.J. Hawke, official opening address, *National Technology Strategy*, *op. cit.*, p 6 (1979 data)
29. Draft discussion, *National Technology Strategy*, *op. cit.*, p 1
30. Australian Academy of Technological Sciences, *Developing High Technology Enterprises for Australia*, AGPS, Canberra, 1983
31. The first seven were: Sydney – BT Innovation Ltd, $10 million, Hambro Grantham (MIC) Ltd, $10 million, and Austech Ventures Ltd, $5 million; Melbourne – Western Pacific Investment Co. Ltd, $10 million, Australian Pacific Technology Ltd, $5 million; Brisbane – Techniche Ltd, $5 million; Perth – Westintech Investment Corp. Ltd, $5 million. See *Australian*, 29.5.84
32. IAC Draft Report, *op. cit.*, Appendix D6
33. *ibid.* Appendix D11-13
34. Helen Meredith, 'Telecom to Foster Local Design Work and Manufacturing', *Australian*, 14.5.84

35. John Costello, 'GEC Wins on Telecom Decision with Videotex', *Australian Financial Review*, 14.6.84
36. *Australian Financial Review*, 28.2.84
37. Contrast chapter 10, 'Technological Determinism: But We Have No Choice' in *Sleepers Wake! Technology and the Future of Work*, OUP, 1982, with the Chairman's Address and Closing Address by B.O. Jones at the National Technology Conference, *op. cit.*, p 10-20 and p 223-228
38. B. Jones, *National Technology Conference*, *op. cit.*, p 10
39. G. Barker, 'Button, An Architect with Grand Design for Industry', *Age*, 2.6.84.
40. B. Jones, National Technology Conference, op cit, p 10

CHAPTER FIVE

1. Henry Mayer, *The Press in Australia*, Lansdowne Press, Melbourne 1964
2. Mark Armstrong, *Broadcasting Law and Policy in Australia*, Butterworths, Sydney, 1982
3. Acknowledgement is made of the adaptation of this chart from 'Outline of Australian Broadcasting', prepared by Andrew Bear in *Mass Media and Modern Society*, Department of Continuing Education, Adelaide University, 1979
4. *Report of The Committee of Inquiry into Telecommunications Services In Australia*, Vols 1-3, AGPS, 1982 (known as the Davidson Report); *The ABC in Review: National Broadcasting in the 1980s*, AGPS Canberra 1981 (known as the Dix Report); *The Report of the Committee of Review of SBS*, AGPS, 1984 (known as the Connor Report)
5. The Australian Broadcasting Tribunal, *Satellite Program Services: Inquiry into the Regulation of the Use of Satellite Program Services by Broadcasters*, AGPS, July 1984, vol 1, p 54
6. Allan Brown, 'Economic Aspects of Cross Media Ownership' in *Australian Journalism Review*, vol 6, no 2, July-December, 1984.
7. Lord Reid in *Ridge* v *Baldwin* (1964) AC64, quoted in Satellite Program Services Report, *op. cit.*, p 176
8. H. Mayer & H. Nelson (eds), *Australian Politics: A Third Reader*, Cheshire, 1972
9. See terms of reference, *Report of the Inquiry into the Ownership and Control of Newspapers in Victoria*, by Hon J.G. Norris, Report to the Premier of Victoria, 15 September 1981
10. Satellite Program Services Report, *op. cit.* p 25
11. *ibid.*, p 25
12. Postal and Telecommunications Department, *Australian Broadcasting: Report on the Structure of the Australian Broadcasting System and Associated Matters*, by F.J. Green, AGPS, 1976, para 126.

13. Section 90C, *Broadcasting and Television Act*

14. ABCB report, para 140, quoted in Satellite Program Services Report, *op cit.*, p 28

15. The licences were granted to Brisbane TV Ltd (BTQ), Queensland Television Ltd (QTL), Southern Television Corporation Pty Ltd (NWS), Television Broadcasters Ltd (ADS)

16. See M. Armstrong, *Broadcasting Law in Australia, op. cit.*, p 144

17. Satellite Program Services Report, *op. cit.*

18. *ibid.*, p 164

19. Data is taken from ABT information on shareholding as at 31 March 1984: See Satellite Program Services Report, *op. cit.*, vol 1, p 292

20. See Norris Report, *op. cit.*, Appendix II.

21. Transcript, ABT Inquiry into Cable and Subscription Television Services and Related Matters, p 1399

22. *ibid.*, p 2349-50

23. R. Whitehead, 'Fairfax Enters New Era with Massive Issue', *Age*, 12.9.84

CHAPTER SIX

1. Adam Smith Institute Omege Report, *Communications Policy*, London, 1984

2. P. Schlesinger, 'Privatising the BBC', *The Listener*, 31.5.84

3. See Alex Brodie, 'Sounding the Death-Knell for the Best in US Radio', in *The Listener*, 16.6.83

4. Employment data as at July 1984 in *ABC Powers and Functions*, an ABC pamphlet

5. Data in SBS submission to Committee of Review, April 1984

6. Data provided by Brendan O'Dwyer, Executive Director, Public Broadcasting Foundation

7. K.S. Inglis, *This is the ABC*, MUP, 1983, p 5

8. *ibid.*, p 25

9. J. Tolhurst, 'Research into Broadcasting to and for Ethnic Communities', in *Media Information Australia*, February 1980, p 53

10. See Third Report of the Ethnic Television Review Panel, *Programming for the Multi Cultural/Multi Lingual Television Service: Objectives and Policies,* AGPS, 1980

11. *Australian Broadcasting: A Report on the Structure of the Australian Broadcasting System and Associated Matters*, Postal and Telecommunications Department, AGPS, September 1976

12. See Richard Harding, *Outside Interference*, Sun Books, 1979, p 93

13. See Harding *op. cit.*, and Inglis, *op. cit.*

14. See Inglis, *op. cit.*, p 389

15. *ibid*, p 420
16. *The ABC in Review: National Broadcasting in the 1980s*, Report of the Committee of Review of the Australian Broadcasting Commission, May 1981, AGPS, vol 1
17. Philip McCarthy and Julie Flynn, 'The A.B.C. After Dix', *National Times*, 14.6.81
18. N.A. Brown, 'Government Decisions in Response to the Recommendations of the Committee of Review of the Australian Broadcasting Commission', Press Release, 4.7.83
19. M. Duffy, 'Labor Acts on A.B.C.' *Australian Financial Review*, 19.4.83
20. The other Board members were Deputy Chairman and Executive Officer of the Australian Family Planning Association, Wendy McCarthy, a former Queensland Liberal Senator, Neville Bonner, a senior lecturer from the University of Western Australia, Sister Patricia Brady, a former IAC Commissioner and economic consultant, Richard Boyer, the ACTU Research Officer and Industrial Advocate, Jan Marsh, and a freelance film and documentary producer, Robert Raymond.
21. See 'Taking a Pillow to Hit the ABC Elephant', in *Australian* (editorial), 2.2.84
22. Dix Inquiry, *op. cit.,* Summary of recommendations, p 3
23. S. Revill, 'Statutory Authorities: Independence, Accountability and the Public Interest', DOC Conference, 17.11.83

CHAPTER SEVEN

1. *Report of the Commission of Inquiry into the Australian Postal Commission* (The Vernon Inquiry), AGPS, Canberra 1974
2. Telecom Australia, *Annual Report* 1982-83
3. *ibid.*, p 14
4. Telecom Annual Reports. See also I. Reinecke & J. Schultz, *The Phone Book*, Penguin Books, 1983, p 16-17
5. *OTC Annual Report*, year ended 31 March 1983, p 4
6. *ibid.* p 2
7. Quoted by G. Noonan, 'Row over OTC Impost', *Australian Financial Review*, 18.11.83
8. Address by Peter Holmes a'Court to the 51st Conference of ANZAAS, Brisbane, 13.5.81
9. Bill Mansfield, federal secretary, ATEA in a 'letter to the editor', *Australian Financial Review*, 24.11.81
10. *Technological Change in Australia: Report*, 4 volumes, AGPS, Canberra, 1980
11. See C.D. Trengove, *Telecommunications in Australia: Competition or Monop-*

oly? Centre for Policy Studies, Special Study no. 4, Monash University 1982. Acknowledgement is also made of J. Langdale, 'Competition in the Australian Telecommunications Industry' in *The Australian Quarterly*, Winter 1982 and J. Langdale, 'Competition in Telecommunications' in *Telecommunications Policy*, December 1982. Also, M. Tyler, 'Regulation and Monopoly in Public Telecommunications: Theory and International Experience' in M.B. Williams (ed), *Pathways to the Information Society*, Proceedings of the Sixth International Conference on Computer Communications, London, 7-10 September, 1982, and in the same volume, P. Hughes, 'Monopoly – Do the Advantages outweigh the disadvantages?'

12. *Report of the Committee of Inquiry into Telecommunications Services in Australia*, Vols 1-3, AGPS, 1982 (The Davidson Inquiry)
13. Submission by Telecom Australia, 'Public Inquiry into Telecommunications Services in Australia', Telecom Australia, February, 1982
14. Business Telecommunications Services Pty Ltd, 'Submission to the Public Inquiry into Telecommunications Services', December, 1981
15. Davidson Report, *op. cit.* vol 1, p 137
16. 'Letter to editor', J. Davidson, Telecom – 'A Need to Emphasize the Facts', *Australian Financial Review*, 3.11.82
17. Davidson Report, *op. cit.*, vol 2, p 36
18. *ibid.*, vol 2, paras 2.31 and 2.32
19. Telecom submission to Davidson Inquiry, *op. cit.*, E.3.7
20. Recommendations 1-7, Davidson Report, *op. cit.*, vol 1, summary, p 28
21. *ibid.*, vol 1, Summary. Recommendation 17, p 29
22. *ibid.*, vol 1, Summary. Recommendation 18, p 29
23. *ibid.*, vol 1, Summary. Recommendation 20, p 30
24. *ibid.*, vol 1, Summary. Recommendation 21, p 30
25. *ibid.*, vol 1, Summary. Recommendation 92, p 40
26. D. Snow, 'Blueprint for Telecom's Reduced Role in Tele-communications', *Australian Financial Review*, 29.10.82
27. N. Brown, press release, 9.2.83
28. *ibid.*
29. See I. Reinecke & J. Schultz, *The Phone Book*, *op. cit.*, p 176-180
30. Ted Knez, 'Unions Condemn Plan for Telecom', *Australian Financial Review*, 29.10.82
31. G. O'Neill, 'Compete More on Telecom', *Herald*, 28.10.82
32. M. Duffy, press release, 20.10.83
33. P. Ellercamp, 'Libs Put Lower Deficit Before Taxation Cuts', *Australian*, 13.4.84
34. D. Solomon, 'Packer's Bid to Open Airwaves', *Australian Financial Review*, 10.8.84

35. H. Meredith, 'ATUG Urges Relaxation of Telecom Institutions', *Australian*, 23.7.84

36. The Australian Federation of Consumer Organisations Inc. Submission to The Prices Surveillance Authority Inquiry into Proposed Charges for Services by Telecom, August 1984

CHAPTER EIGHT

1. Report and Ministerial statement, National Communications Satellite Task Force, A. Staley, House of Representatives, Hansard, 18.10.79 p 2224
2. *ibid.*, p 2224-5
3. Quoted by Ann Moyal, *Clear Across Australia*, Nelson, 1984, p 341
4. A. Staley, *Domsat '79: A Communications Satellite for Australia*, Centre for Continuing Education, ANU, Canberra, 1979, p 18
5. Ministerial Statement, Ian Sinclair, Minister for Communications, 6.5.82
6. Deborah Snow, 'Aussat Fires First Broadside at Telecom', *Australian Financial Review*, 28.9.82
7. G. Gosewinckel, *AM* and *The World Today*, ABC Radio, 17.6.83
8. According to *Intelsat: Annual Report* in the year 1981-2 demand for television programming time increased 131 per cent over previous Australian demand in the 1980-81 year.
9. R. Stuart of New York Times in 'Intelsat Brought Down to Earth', *Australian Financial Review* 13.11.84
10. *Intelsat Annual Report*, 1982
11. *Video Age*, February 1983, p 6
12. R. Milliken, 'Packer Plans a Satellite Network', *National Times*, 4.11.78, p. 25
13. *Video Age*, February 1983, p 7
14. Donald S. Bond, *The Opportunity for Television Program Distribution in Australia Using Earth Stations*, Television Corporation Ltd, August 1977, p 51
15. *ibid*, p 53
16. *National Communications Satellite System*, Commonwealth Task Force Report, July 1978, p xiv
17. *ibid.*, p xiv
18. *ibid.*, p 114
19. *ibid.*, p 116
20. *ibid.*, p 116
21. *ibid.*, p 114
22. *National Communications Satellite System*, Working Group Report, AGPS, August 1979, p xi
23. *ibid.*, p 189
24. *ibid.*, p 183

25. *ibid.*, p xiii

26. *ibid.*, p xiv

27. *ibid.*, p xv

28. A. Staley, Communications Satellite Task Force, House of Representatives, *Hansard*, p 2224-2229, and the *Age* 20.9.79 and 18.10.79

29. *ibid.*, p 2226

30. The Board of Aussat Pty Ltd consisted of the following: chairman: E.S. Owens, director of a number of Australian companies; deputy chairman: Dr D.D. Brown, formerly deputy general manager of CSR; H.S. Cottee, a Commissioner of the Overseas Telecommunications Commission (Australia); J.J. D'Arcy, financial director, Queensland Press Ltd; T. Farrell, assistant general manager, John Fairfax Ltd; J.F. Gleeson, chairman and managing director, Telecasters North Queensland Ltd; R. Graham, a director of Australian Consolidated Press, Publishing and Broadcasting Ltd, and General Television Corporation Ltd; Professor M.W. Gunn, Dean of the Faculty of Engineering, University of Queensland; D.B. Horgan, chairman of Metro Industries; Mr E.E. Payne, deputy secretary, Department of Communications, and a Commissioner of OTC; R.H. Searby, Q.C., chairman, News Ltd; R.H. Treweeke, federal president Isolated Children's Parents Assoc.; R.F. Weisener, joint managing director Aviation Developments Pty Ltd

31. *Australian Financial Review* (editorial), 7.7.83

32. Margot O'Neil, 'Cabinet Gives Aussat Satellite the Go Ahead', *Age*, 14.7.83

33. Bill Birnbauer, 'Satellite Share Sale Unlikely, Says Duffy', *Age*, 20.6.83

34. Deborah Snow, 'Telecom Puts Firm Case to Control Aussat', *Australian Financial Review*, 22.6.83

35. The outgoing directors represent all major interests in commercial television: John D'arcy from Queensland Press, Thomas Farrell from John Fairfax and Sons, John Gleeson from Regional Television in Queensland, Robert Graham of Kerry Packer's Publishing and Broadcasting, and Richard Searby of News Ltd. The new appointees: Sydney merchant banker, David Hoare as chairman, and as directors, John Garden (executive director of CRA Ltd.), William Dix (managing director, Ford Motor Co.), Professor Helen Hughes (Australian National University), Reginald Robin (formerly general manager and commissioner of the Australian National Line), and Mr Bill Pollock (managing director, Telecom)

36. 'Development of Satellite Related Broadcasting Services Ownership of the Satellite System', Ministerial Statement, M.J. Duffy, Minister for Communications, 15.11.83

37. Richard McGregor, 'The Outback Joins the Global Village', *Sydney Morning Herald*, 16.11.83

38. Ministerial Statement, 15.11.83

39. 'A Slice of the Pie in the Sky', *Four Corners*, ABV2, 10.9.83

CHAPTER NINE

1. *Aussat – How Satellites Work*, Aussat Pty Ltd, Sydney 1983, promotional literature

2. Ministerial Statement, Ian Sinclair, Minister for Communications, 6.5.82

3. Ministerial Statement, Michael Duffy, Minister for Communications, 15.11.83

4. 'New satellites at Telecom 83', *Interavia* no. 1 1984

5. *New Technology in the Media: The Domestic Satellite*, Basic Paper no. 5, Legislative Research Service, Department of the Parliamentary Library, Commonwealth of Australia, 1982, p 65

6. *ibid.*, p 65

7. *The ABC in Review: National Broadcasting in the 1980s*, (The Dix Report) vol 1, AGPS, May 1981, ref 27.19

8. Acknowledgement is made of articles by Don Knox, chief communications engineer, Air Traffic Systems Branch, Department of Aviation of 'Aussat to Include Provision for Aviation Communications', in *ICAO Bulletin*, December 1982, and 'Aussat Boom to Aviation' in *Asian Navigation*, August 1982. Also to Robert Bell, 'Aussat '85: It'll be a VHF Revolution' in *Australian Aviation*, March 1983

9. *Remote Area Telecommunications Study*, National Report, Telecom Australia, August 1980, p 17

10. *A Study of Remote Area Telecommunications in the Northern Territory*, vol 1 Summary Report, IMG Group Pty Ltd, May 1980, p 50

11. M.J. Duffy, Minister for Communications, Press Release, 12.10.83

12. *ibid.*

13. B. Blundell, 'How AAP-Reuters Delivers Information', speech to Librarians Association, Perth, 29.9.83

14. AAP submission to Labor Government's Caucus sub-committee on Communications, chaired by J. Reeves (NT) and D. Charles (Vic), August 1983

15. R.B. Lansdowne, 'National Communications Satellite System', speech to AANA, Sydney, 2.3.80

16. Task Force on Broadcasting Development, *Satellite Broadcasting Development Project*, Discussion Paper no. 1, Department of Communications, Canberra, November 1981

17. Australian Broadcasting Tribunal, *Satellite Program Services*: Inquiry Into Regulation of the Use of Satellite Program Services by Broadcasters. Report July 1984, AGPS vol 2, p 205

18. *Domsat 79*, Papers and Proceedings of the first national domestic satellite

conference, Centre for Continuing Education, Australian National University, 1979, p 89

19. Ministerial statement, M.J. Duffy, 'Development of Satellite Related Broadcasting Services', 15.11.83

20. *ibid.*

21. Satellite Program Services Report, *op. cit.*, vol 1 p 280 ref 4.117

22. *ibid.*, vol 1, p 77

23. *ibid.*

24. *ibid.* vol 1, p 79

25. M.J. Duffy, press release, 2.10.84

26. Satellite Program Services Report, op cit., vol 1, p 55

27. *Localism In Australian Broadcasting: A Review of the Policy* DOC Canberra, 1984 p 174 (The Oswin Inquiry)

28. Data supplied in published annual advertising data reports of CEASA, the Commercial Economic Advisory Service of Australia, Sydney

29. *ibid.*

30. The top seven were (in order) Rockhampton TV, Mackay TV, Wide Bay TV, Country TV, TV New England, Darling Downs TV, Riverina TV. See *Australian Business*, 9.5.84, p 68

31. Satellite Program Services Report, vol 1, p 340

32. *ibid.*

33. M.J. Duffy, press release, 10.10.84

34. *ibid.*

35. Satellite Program Services Report, op cit, vol 2, p 2

36. *Domsat 79*, ANU, Canberra, 1979, p 19

37. *PM*, ABC Radio, 6.5.82

38. Aussat: *Australia's National Satellite System*, published by Aussat Pty Ltd, June 1983, p 23

39. *Policy Issues and Public Sector Costs: A Study into An Australian Domestic Communications Satellite System: Aussat*, published by the ATEA, June, 1983

40. *ibid.*

41. C. Cooper, 'letter to editor', 'The A.T.E.A. was Conservative', *Australian Financial Review*, 13.7.83

42. M.J. Duffy 'A.B.C. Earth Station Installation Program Announced', press release, 21.8.84

43. 'Aviation Satellite System to Go Ahead', reporting Minister for Aviation, Mr Fife, *Australian*, 25.8.82

44. *Policy Issues and Public Sector Costs*, ATEA, p 1

45. M. O'Neil, 'Cabinet Gives Aussat Satellite the Go-Ahead', *Age*, 14.7.83

46. See Tony Boyd, 'Strong Demand for Aussat System Will Force Early Launch of Third Satellite', *Australian*, 11.5.84
47. 'Arianspace Wins $24m Contract To Send Satellite', *Australian*, 15.10.84
48. Don Kennedy, at the Law Institute of Victoria, 5.8.83, gave 80 per cent as the figure and David Major, at a Corporate Communications Seminar 15.8.83, Melbourne, gave 50 per cent as the figure. Both were senior staff members of Aussat Pty Ltd.
49. Satellite Program Services, *op. cit.*, vol 2, p 1
50. *ibid.* vol 2, p 1
51. *ibd.* vol 2, p 76
52. House of Representatives, Hansard, 8.3.84, p 697

CHAPTER TEN

1. *Interfutures: Facing the Future, Mastering the Probably and Managing the Unpredictable*, OECD, Paris, 1979
2. *Privacy*, Law Reform Commission Report no 22, volume 1, AGPS, 1983, p 238
3. *ibid.*, p 69
4. *ibid.*, p 57
5. Alvin Toffler, *The Third Wave*, Kroft, New York, 1980
6. Anthony Smith, *Goodbye Gutenberg: The Newspaper Revolution of the 1980s*, OUP New York, 1980, p 22
7. Ithiel de Soia Pool, *Technologies of Freedom* Harvard University Press, Massachusetts, 1983
8. For cable television penetration see T. Hollins, *op. cit.*, below; for subscription television see ABT, *Cable And Subscription Television Services for Australia*, AGPS 1982, vol 1 page 4.28; for home computers see 'Publishers Go Electronic' in *Business Week* 11.6.84; for video cassette recorders see *Inter-Media*, vol 11 September 1983, p 73
9. See K. Windchuttle, *The Media*, Penguin, 1983, pp 22-39
10. A Brown 'The Australian Newspaper Industry', Seventh Conference of Economists, Macquarie University, 1978, quoted in C Stuart, "Video text Poses Financial Threat to Newspapers" in *Australian Journalism Review*, vol 5, no 1, January 1983
11. H. Gordon, 'The Future of Newspapers', *Quadrant* January 1977
12. H. Wilson & W. Bonney, *Australia's Commercial Media* Macmillan, 1982
13. G. Taylor, 'Newspapers and New Technology', 3AW Seminar, Melbourne 26.5.83
14. A. Smith, 'Information Technology and the Age of Abundance' in *Daedalus*,

Fall 82, p 11
15. *ibid.*, p 17
16. Peter Furze, 'The Traditional Information Supplier In the New Communications Age' RMIT Conference 1.7.84
17. Kern, Finn, 'Advertising: Facing the Threat From the New Media', *FIEJ Bulletin*, May 1981 no 126, quoted in Stuart, *op. cit.*
18. Stuart, *op cit.*, p 45
19. 'The Computer Moves In', *Time* 3.1.83
20. 'The Coming Shake Out In pcs', *Business Week*, 22.11.82
21. 'The Godzilla of Armonk', *Australian Financial Review*, 5.1.84 (republished from the *New York Times*).
22. See G. Gerlach, *Business Review Weekly*, 13.2.82, and 'How the VCR is Changing Australia' in *Bulletin*, 5.10.82 and D. Elias, 'Video' *Age* 3.11.84
23. T. Whitmont, 'TV Hire At Crossroads As VCR Market Peaking', *Australian Financial Review* 12.7.84
24. See D. Shires, 'VCRs Have Minimal Effect on Viewing, Survey Claims', *Australian Financial Review* 17.4.84, also J. Munton, "VCR Growth No Threat to Advertisers, Study Shows", *Australian Financial Review* 3.7.84
25. J. Munton, 'High Domestic VCR Usage A Problem for Advertisers' in *Australian Financial Review* 6.3.84
26. 'How TV Is Revolutionising Hollywood', *Business Week*, 21.2.83, p 42
27. T. Hollins, *Beyond Broadcasting: Into The Cable Age*, BFI Books, 1984, p 133
28. 'The Gold Rush of 1980', *Broadcasting*, 31.3.80
29. T. Hollins, *op. cit.,* p 125
30. 'Cable TV: The End of The Shake Out', in *Financial World*, 2.5.84, p 15
31. 'Cable Television – Survey of the Community Channel', CRTC March 1979
32. See Hollins *op. cit.*, p 138 and P. Taylor, 'U.S. Cable Industry Comes Down To Earth', *Australian Financial Review* 23.12.83
33. *Financial World*, 2.5.84, p 15
34. The Australian Broadcasting Tribunal, *Cable and Subscription Television Services for Australia* AGPS August 1982
35. *ibid.*, p 64
36. *ibid.* Main Conclusions and Recommendations, p 85
37. *Report of the Committee of Inquiry Into Telecommunications Services In Australia*, AGPS Canberra 1982
38. ACASA Conference, Canberra, 9.12.82 and in an interview with T. Krause, *Australian*, 4.3.83
39. Transcript of ABT Cable Television Inquiry, p 1715-1716a, 20.11.81

CHAPTER ELEVEN

1. W.H.G. Armytage, 'The Rise of The Technocratic Class' in D.O. Edge and

J.N. Wolfe (eds), *Meaning And Control*, Tavistock, London, 1973, p 65

2. *The Australian Broadcasting Tribunal, Satellite Program Services: Inquiry Into The Regulation of The Use of Satellite Program Services By Broadcasters.* Report volumes 1 and 2 AGPS July 1984

3. Quoted by Michael Tiger, 'Courting The Cost of Information', *Inter-Media* vol 12 no 2 March 1984

4. *ALP, 1982 Platform Construction and Rules*, p 147-154

5. S. Nora and A. Minc, *op. cit.*

6. Satellite Program Services Report, *op. cit.*, vol 1, p 86

7. *ibid.*, vol 1, p 19

8. *ibid.*, vol 1, p 109-123

SELECT BIBLIOGRAPHY

Armstrong, Mark, *Broadcasting Law and Policy in Australia*, Butterworths, Sydney, 1982.

New Media: Law and Policy, proceedings of a seminar held August 1981 by the Faculty of Law, University of NSW, in conjunction with the Australasian Communications Law Association, Sydney.

Aussat: *Australia's National Satellite System*, Aussat Pty Ltd, June 1983.

Barr, Trevor, *Reflections of Reality: The Media In Australia*, Rigby, Adelaide, 1977.

Bear, Andrew, *Public Broadcasting In Australia*, Studies In Continuing Education, University of Adelaide, 1979.

The Emergence of Public Broadcasting In Australia, Australian Journal of Communication, No. 4, July 1983.

Bell, Daniel, *The Coming of Post Industrial Society*, Heinemann Educational Books, London, 1974.

Boehringer, K., Bonney B., Rowse T., and Wilson H. (eds), *Media in Crisis*, proceedings of the third Communication Technology and Control Conference, NSW Institute of Technology, 22-25 August 1980, Broadway, NSW, Communication, Technology and Control Foundation, 1980.

Bond, D.S. *The Opportunity For Televised Programme Distribution in Australia Using Earth Stations*, Television Corporation Ltd, Sydney, 1977. (The Bond Report)

Botkin J., Dimancescu D., Stata R., *Global Stakes: The Future of High Technology In America*, Harper & Row, New York, 1982.

Broadcasting In Australia: Today's Issues And The Future, proceedings of the National Conference, July 1980 at Australian National University, Centre For Continuing Education, ANU, 1980.

Broadcasting: The 50th Anniversary Issue, Vol. 101 No. 15, Washington, 1981.

Brett, Judith, *Planning The Future: Rhetoric And Phantasy In Telecom 2000*, in Melbourne Journal of Politics, 12, 1980.

Carlyon, Les, *Paperchase: The Press Under Examination*, Herald And Weekly Times, Melbourne, 1982.

Dertouzos M. and Moses J., *The Computer Age: A Twenty Year View*, MIT Press, London, 1980.

Dizard, Wilson P., *The Coming Information Age*, Longman, New York, 1982.

Domsat '79: A Communications Satellite for Australia, Centre for Continuing Education, ANU, Canberra, 1979.

Dugdale, Joan, *Radio Power: A History of 3ZZ*, Hyland House, Melbourne, 1979.

Duke, Chris, *The Impact of Modern Communication Technology in Australia*, Canberra, Centre for Continuing Education, ANU, 1979.

Edgar, P.M. and Rahim, S. (eds), *Communication Policy in Developed Countries*, Routledge & Kegan Paul, 1983.

Evans, Christopher, *The Mighty Micro*, Coronet, 1979.

Forester, Tom (ed), *The Micro-electronics Revolution*, Blackwell, London, 1980.

Godfrey D. and Parkhill D., *Gutenberg Two*, Press Porcepic Ltd, Toronto, 1980. (Revised edition).

Habermas, Jurgen, *Legitimation Crisis*, Heinemann, London, 1976.

Towards a Rational Society: Student Protest, Science and Politics, translated by Jeremy J. Shapiro, Heinemann, London, 1971.

Harding, R.W., *Outside Interference: The Politics of Australian Broadcasting*, Sun Books, Melbourne, 1979.

Hollins, T., *Beyond Broadcasting: Into The Cable Age*, BFI, 1984.

Home Video Yearbook, 1981-2, Knowledge Industry Publications, New York, 1982.

Howkins, *New Technologies, New Policies,* BFI, London 1982.

Inglis, K.S. *This Is The ABC*, MUP, Melbourne, 1983.

Jones, Barry, *Sleepers, Wake! Technology and the Future of Work*, OUP, Melbourne, 1982.

Jonscher, Charles, 'The Economic Causes of Information Growth', *Intermedia*, Nov 1982.

Kirby, Michael, *Reform The Law*, OUP, Melbourne, 1983.

The Law And Modern Technology, Deakin University Press, 1982.

Langdale, John, *International Perspectives on Competition in the Telecommunications Industry: A Critique*, in Australia's Industrial Future: Communication, ANZAAS.

Competition in Telecommunications, Telecommunications Policy, December, 1982, pp 283-299.

Government Policy and the Information Economy, Media Information Australia, No 28, May 1983.

Large, Peter, *The Micro Revolution*, Fontana, 1980.

MacBride, Sean (ed), *Many Voices, One World*, Report by The International Commission for the Study of Communication Problems, UNESCO, 1980.

Martin, James, *Telematic Society, A Challenge For Tomorrow*, Prentice Hall, 1981.

Future Developments In Telecommunications, Second Edition, Prentice Hall, 1977.

Mattelart, Armand, *Multinational Corporations and the Control of Culture*, Translated from the French by Michael Chanan, Harvester Press, Sussex, 1979.

McQueen, H., *Australia's Media Monopolies*, Widescope, Melbourne, 1977.

Moss, Mitchell M., *Telecommunications And Productivity*, Addison Wesley, Mass., 1981.

Moyal, Ann, *Clear Across Australia*, Nelson, Melbourne, 1984.

New Information Technology: Impacts On Families In Australia, An annotated bibliography compiled by Ruth Beresford, Institute of Family Studies, Melbourne, 1983.

Nora, Simon & Minc, Alain, *The Computerisation of Information: A Report to the President of France*, The MIT Press, Cambridge, Mass., 1980.

Pawley, Martin, *The Private Future*, Thames and Hudson, London, 1973.

Pelton, J.N., *Global Talk*, Sijthoff & Noordhoff, 1981.

Pool, Ithiel de Sola, *Technologies of Freedom*, Harvard University Press, Mass., 1983.

Reinecke, Ian, *Micro Computers*, Penguin, Melbourne, 1983.

Micro Invaders, Penguin, 1982.

Reinecke, I. & Schultz, J., *The Phone Book*, Penguin, 1983.

Robinson, Glen O. (ed), *Communications for Tomorrow: Policy Perspectives for the 1980's*, New York, Praeger Publishers, 1978.

Salavaggio, Jerry L., *Telecommunications Issues and Choices for Society*, Longman, 1983.

Schiller, Herbert I., *Who Knows: Information in the Age of the Fortune 500*, Ablex, New Jersey, 1981.

Smith, Anthony, *Goodbye Gutenberg*, OUP, 1980.

Information Technology the Myth of Abundance, Daedalus, Fall, 1982.

Sommerlad, E. Lloyd, *The Satellite Dilemma*, in Current Affairs Bulletin, May 1982.

Souter, G., *Company of Heralds: a Century and a half of Australian Publishing by John Fairfax and its Predecessors*, MUP, Melbourne, 1981.

Stonier, Tom, *The Wealth of Information*, Thames Methuen, London, 1983.

Technological Change: Impact of Information Technology 1981-4, AGPS, Canberra, 1983.

Toffler, Alvin, *The Third Wave*, Collins, 1980.

Trengrove, *Telecommunications in Australia: Competition or Monopoly?* Centre for Policy Studies, Special Study No. 4, Monash University, 1982.

Von Bibra, Henry, 'Legal Ramifications for Implementing Cable Television in Australia', in M. Armstrong (ed), *New Media: Law & Policy*, University of NSW, 1982.

Webster, F., & Robins, K., 'Mass Communications And "Information Technology"', in R. Miliband & J. Saville (eds), *The Socialist Register*, 1979, Garden City Press, London, 1979.

Weeramantry, C.G., *The Slumbering Sentinels: Law And Human Rights, in The Wake of Technology*, Penguin, Melbourne, 1983.

Wicklein, John, *The Electronic Nightmare: The New Communications and Freedom*, Viking, 1981.

Williams, Raymond, *Television: Technology And Cultural Form*, Fontana, London, 1974.

Wilson, P.D., *The Coming Information Age: An Overview of its Technology, Economics and Politics*, Longman, New York, 1982.

Windchuttle, K., *The Media*, Penguin, 1983.

MAJOR OFFICIAL REPORTS

Australian Academy of Technological Sciences, *Developing High Technology Enterprises for Australia*, AGPS, Canberra, 1983.

Australian Broadcasting Tribunal, *Cable and Subscription Television Services for Australia*, AGPS, August 1982. (The Jones Report).

Australian Broadcasting Tribunal, *Satellite Program Services: Inquiry into the Regulation of the use of Satellite Program Services by Broadcasters*, AGPS, July 1984.

Australian Telecommunications Commission, National Telecommunications Planning (NTP) Branch, *Telecom 2000: an exploration of the long-term development of telecommunications in Australia*, Melbourne, Telecom Australia, 1975.

Australian Telecommunications Commission, National Telecommunications Planning (NTP) Branch, *Outcomes from the Telecom 2000 report*, Melbourne, Telecom Australia, 1978

Committee of Review of the Special Broadcasting Service, *Serving Multi-Cultural Australia: The Role of Broadcasting*, AGPS, 1985. (The Connor Report).

Committee of Inquiry into the Monopoly Position of the Australian Postal Commission: Report, AGPS, Canberra, August 1982. (The Bradley Report).

Committee of Inquiry into Technological Change in Australia, *Technological change in Australia*, report of the Committee of Inquiry into Technological Change in Australia, Canberra, AGPS, 1980. (The Myers Report).

Commonwealth Government Task Force on the National Communications Satellite System: Report, AGPS, Canberra, 1978.

Commonwealth Government Task Force, *National Communications Satellite System*, Report July 1978, AGPS, 1978. (The White Task Force).

Industries Assistance Commission, *Report on Computer Hardware and Software; Typewriters, Calculating and other Office Machines; Prints and Accessories; Recording Media*, AGPS, 1984.

Localism In Australia Broadcasting: A Review of the Policy, DOC, Canberra, 1984. (The Oswin Inquiry).

Law Reform Commission, *Privacy*, Report No. 22, AGPS, 1983.

National Technology Conference, Proceedings and Report of the Conference held in Canberra, 26-28 September 1983, Department of Science and Technology, AGPS, Canberra 1983.

National Technology Strategy, Discussion Draft, National Technology Conference, Department of Science and Technology, August 1984.

Postal and Telecommunications Department, *Australian Broadcasting: Report on the Structure of the Australian Broadcasting system and Associated Matters*, by F.J. Green, AGPS, 1976. (The Green Report).

Report of the Committee of Inquiry into Telecommunications Services in Australia, Vols 1-3, AGPS, 1982. (The Davidson Inquiry).

Report of the Commission of Inquiry into the Australian Postal Commission, AGPS, Canberra, 1974. (The Vernon Inquiry).

Report of the Inquiry into the Ownership and Control of Newspapers in Victoria, by Hon J.G. Norris, Report to the Premier of Victoria, 15 September 1981.

Report to the Department by W.D. Scott & Co Pty Ltd, in association with Arthur D. Little Inc, *Information Technology in Australia*, Department of Science and Technology, July 1984.

Report to the Prime Minister by The Australian Science and Technology Council (ASTEC), *Technological Change And Employment*, AGPS, June 1983.

Satellite Broadcasting Development Project, Department of Communications, Discussion Paper 1, November 1981.

Task Force on Broadcasting Development, *Satellite Broadcasting Development Project*, Discussion Paper No. 1, Department of Communications, Canberra, November 1981.

Telecommunications: Pressures And Policies For Change, OECD, Paris 1983.

The Australian Electronics Industry, A Report by the Electronics Industry Advisory Council, AGPS, 1980.

The ABC in Review: National Broadcasting in the 1980s, Report of the Committee of Review of the Australian Broadcasting Commission, May 1981, AGPS. (The Dix Report).

New Technology in the Media: The Domestic Satellite, Basic Paper No. 5, Legislative Research Service, Department of the Parliamentary Library, Commonwealth of Australia, 1982.

Working Group Report, *National Communication Satellites System*. Report August 1979, AGPS, 1979.

Index